JAWS

JAWS

THE STORY OF A HIDDEN EPIDEMIC

SANDRA KAHN and PAUL R. EHRLICH

Foreword by ROBERT SAPOLSKY

AN **ENVIRONMENTAL HEALTH SCIENCES** BOOK

STANFORD UNIVERSITY PRESS • STANFORD, CALIFORNIA

Stanford University Press
Stanford, California

Illustrations by Susan Szecsi

An Environmental Health Sciences Book

Printed in the United States of America on acid-free, archival-quality paper

ISBN 9781503613584
First paperback printing, 2021

The Library of Congress has cataloged the hardcover edition as follows:

Names: Kahn, Sandra, Dr., author. | Ehrlich, Paul R., author.
Title: Jaws : the story of a hidden epidemic / Sandra Kahn and Paul R. Ehrlich.
Description: Stanford, California : Stanford University Press, 2018. |
  Includes bibliographical references and index.
Identifiers: LCCN 2017058481 (print) | LCCN 2017058937 (ebook) |
  ISBN 9781503606463 (e-book) | ISBN 9781503604131 (cloth : alk. paper)
Subjects: LCSH: Jaws—Health aspects. | Jaws—Abnormalities. | Malocclusion.
Classification: LCC QM105 (ebook) | LCC QM105 .K34 2018 (print) |
  DDC 611/.92—dc23
LC record available at https://lccn.loc.gov/2017058481

Cover design: Rob Ehle
Cover image: Dr. María José Muñoz

*To John and Mike Mew
in recognition of their great service to humanity,
and to David, Ilan, Ariela, and Anne
for their patience and support.*

CONTENTS

FOREWORD

The Surrealist painters were fond of an epigram penned by an obscure 19th century French poet. "Beauty," they would say, is the "chance meeting on a dissecting table of a sewing machine and an umbrella." This was a celebration of the Surrealist's love of random, capricious events; of absurd, dislocating juxtapositions. The book that you are holding generates a different sort of epigram—"intensely interesting," it suggests, can be the outcome of the "chance meeting over a dinner table of an orthodontic scholar and an eminent evolutionist."

Human cultural evolution has been one long string of examples of the law of unexpected consequences. We invent agriculture, which leads to food surpluses, which leads to job specialization, and before you know it, we've invented socioeconomic status, the most crushing way of subordinating the low ranking that primates have ever seen. We invent sedentary dwelling and permanent structures, and soon we're dealing with the public health consequences of something no self-respecting primate would ever do—living in high-density populations in close proximity to its feces. We domesticate wolves into being companions, and soon we're dressing up our dogs in Halloween costumes and buying Pet Rocks. The emergence of modern humans has generated some surprising twists and turns.

Kahn and Ehrlich explore one of these unexpected consequences of human culture, sitting at the intersection of the expertise of this unlikely pairing of authors. Who would ever have predicted that the Agricultural Revolution, the Industrial Revolution, and the Westernization of nursing patterns would have led to a distinctive orthodontic profile (in both the metaphorical and literal sense of "profile")? And who would have predicted that this orthodontic profile winds up being relevant to an array of aspects of child development, health, and disease? Most of all, who would have predicted that such a capricious combination of authors could have produced a book both extremely interesting and extremely important? If you have kids, like kids, were ever a kid, or have a jaw, it's well worth your while to read.

Robert Sapolsky
Neuroscientist and author,
Why Zebras Don't Get Ulcers, and *A Primate's Memoir*
Stanford University

ACKNOWLEDGMENTS

David Leventhal and Anne Ehrlich suffered more with this book than we can ever tell. The only person who suffered more was Paul's (and now Sandra's) good friend and frequent editor, Jonathan Cobb. His work on the manuscript transformed *Jaws*, making it orders of magnitude better. An anonymous reviewer at Stanford Press made many helpful suggestions.

Ellyn Bush, Richard Klein, John Mew, Mike Mew, and Simon Wong were very helpful answering questions along the way. A bunch of other friends and colleagues took time from their busy lives to read and comment on the entire manuscript or critical parts of it. For their enormous help we are indebted to Andy Beattie, Keira Beattie, Margaret Bergen, Corey Bradshaw, Greg Bratman, Kate Brauman, Marie Cohen, Gretchen Daily, Lisa Daniel, Joan Diamond, Jared Diamond, Nadia Diamond-Smith, Anne Ehrlich, Jeremy Feldman, Marc Feldman, Daniel Friedman, John Harte, Mel Harte, Craig Heller, Jill Holdren, David Leventhal, Simon Levin, Karen Levy, Jess Marden, Chase Mendenhall, John Morris, Pete Myers, Graham Pyke, Barry Raphael, Robert Sapolsky, John Schroeder, Susan Thomas, Chris Turnbull, and Kenneth Weiss.

Alan Harvey and his colleagues at Stanford University Press aided us in many ways, as did our agent, Jim Levine. Margaret Pinette did a magnificent job of copy-editing. It's a great pleasure to work with real professionals.

Subsequent to the appearance of the first edition of Jaws, we have published a peer-reviewed scientific article updating the message of the book: Kahn S, Ehrlich P, Feldman M, Sapolsky R, Wong S. 2020. The jaw epidemic: Recognition, origins, cures, and prevention. BioScience 70:759–771. We are very grateful to the scientists who joined us in that enterprise.

JAWS

INTRODUCTION

This is a story about a vast and serious epidemic afflicting the developed world increasingly over the last few centuries, one that has gone virtually unrecognized. *Jaws* is about its origins, how it was discovered, and what we can do about it. The epidemic's roots lie in cultural shifts in important daily actions we seldom think about; we just do them automatically. We don't think about chewing, breathing, growing, or sleeping, or even the position of our jaws when we're not eating or talking. Most of these actions we don't acquire as habits, that is, by doing them repeatedly; they are inborn. A newborn exposed to air starts to breathe and cry. A baby presented with a nipple opens her mouth, starts to suckle, and after a bit may reward you with a grin. In the evening, after driving you nuts with screaming, your baby sleeps like a log, no training required.

Simple and normal actions, yes. But, we argue, if repeatedly done in certain ways, early in life especially, over time they can undermine your health and alter your appearance in some surprising ways. If you keep your jaws apart and breathe through your mouth rather than through your nose for a few days, bite your tongue once in a while, or have insomnia for a few nights, you are going to be just fine. On the other hand, if you from an early age develop the habits of perpetually mouth breathing, eating mostly soft foods that require little chewing, and sleeping restlessly, snoring and squirming through every night, that could lead to distorted

Image 1. From an early age, babies can get into the habits of mouth breathing.

development of your jaws, face, and airway (the passage through which air enters and leaves the lungs) and to serious health problems later on—even to an early death. You would be a victim of a growing epidemic.

Modern industrialized societies are plagued by small jaws and crowded, ill-aligned teeth, a condition that the dental profession refers to as "malocclusion" (literally "bad bite"). Malocclusion is often accompanied by mouth breathing. Together, not to mention their negative effects on appearance, the two tend to reduce our quality of life and make us more susceptible to disease. And they are increasingly common. William Proffit, author of the most widely used textbook in orthodontics, the part of dentistry focused on straightening crooked teeth, pointed out the scale of the epidemic in the United States in 1998: "Survey data reveals that about a fifth of the population has significant malocclusion, and irregularity in the incisors (crowding of the front teeth) is severe enough in 15% that both social acceptability and function could be affected. Well over half have at least some degree of orthodontic treatment need."[1] A study of people in Sweden in 2007 showed that about a third of the population was in "real need" of orthodontic treatment and almost two-thirds has real or "borderline" need.[2] Orthodontist and clinical director of the London School of Facial Orthotropics, Dr. Michael Mew, asserts that 95 percent of modern humans have deviations in dental alignment; 30+ percent are recommended to have orthodontic treatment (half have extractions); and 50 percent have wisdom teeth removed.[3] If industrialized societies are plagued by jaw problems, might it not be smart to consider what changes might be made in those societies to ameliorate the problems?

Image 2. Proper facial structure and posture. This young man has had a very active life with minimal processed foods. He currently has all his teeth, including wisdom teeth, and did not need orthodontic treatment. (Photo by Steven Green.)

The focus of almost all orthodontic practitioners today is crooked teeth, the straightening of which is the bread and butter of the orthodontic trade. But it may be that most orthodontists are concerned with the least of the jaw-related problems. Crooked teeth, other than their impact on appearance, are virtually harmless. Crooked teeth are, however, a signal of a more basic problem, poor development of the jaws. And distorted jaws influence more vital functions. For example, more than 10 percent of children may now have jaw-related potentially dangerous interrupted breathing at night;[4] in one study in an urban area of Brazil, 55 percent of 23,596 children aged 3 to 9 years were mouth breathers.[5] Although there has not been a coordinated effort to systematically gather data on the frequency of malocclusion, mouth breathing, sleep disturbance and the like, wherever they *are* examined they turn out to be common. Consider: if just 10 percent of the people in the United States were in bed with the flu, all the mass media would be focused on the "flu epidemic."

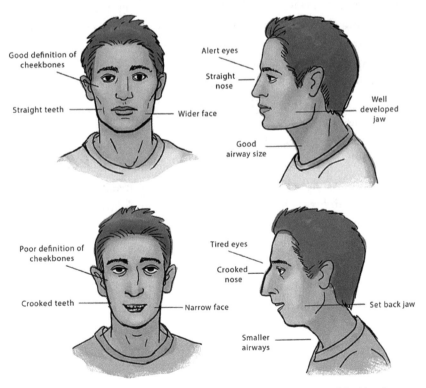

Image 3. Basic differences in facial development of (top) a nose- and (bottom) a mouth-breather.

By now you may be asking yourself, "Who are these people telling me there is a vast public health epidemic that is being ignored? Who is claiming that a long-admired profession may not be paying enough attention to a serious problem in its area of specialization? Who has the chutzpah to proclaim a need to dramatically change some basic aspects of industrial society?" Is this a standard "eat a pound of radishes a day and live a decade longer while enjoying a better sex life" kind of book? Actually, it isn't. It is the result of a collaboration between two concerned scientists with very different backgrounds and experiences—a highly experienced dental practitioner and a world-recognized environmental scientist and expert in human evolution. And we are not selling any product or service.[6]

So how did these two scientists decide to write a book about this unrecognized epidemic? It started as a dinner club; Sandra and Paul and our respective partners, David and Anne, would meet for dinner in Palo Alto at one of several quality establishments every few weeks. The goal was to enjoy some good wine, good food, and good conversation about nature conservation, about how the world was a mess, and to wonder whether it was too far gone to save. It was during these dinners that Sandra started recounting to Paul and Anne a personal journey in her profession as an orthodontist. It was such a striking story, and of so much interest to Paul, that it culminated in his suggesting that they should write a book about it together. Sandra couldn't believe that someone as published as Paul (with more than 50 books and 1,000 articles to his credit) would be interested in her work, but it was exactly her work that he found so interesting, the fact that something so life changing and dangerous was literally right under our noses and we didn't see it. Paul had written a book or two on the same sort of life-changing issues, such as reproduction and racism, but this would be the first one that looked at such an issue from the fresh viewpoint that Sandra brought to the table.

Unlike Paul, who has three grandchildren, Sandra has two young children of her own. And as an orthodontist who had practiced the craft for 22 years, she discovered that she could not treat her own children the same way she was treating her other patients. She realized that, as in so many other professions, dental schools were pumping out students whose practices were "by the book" but were not necessarily best for patients. What she saw in practicing orthodontics the traditional way was that the solution to fixing smiles was usually to extract teeth, wire up the remaining teeth and use the resulting extra space to create beautiful smiles. And the results were exactly that and only that, beautiful smiles. But the smiles lacked "context." These were smiles that in the process of building up a straight set of perfectly aligned pearly whites, left behind destruction to what could have been a strong jaw line, easy breathing, and a well-constructed face. Faces and health were left behind in the race to create that perfect movie-star smile.

So when Sandra was looking for the right way to treat her eldest child without extracting teeth, she first turned to "myofunctional therapy" as a rising and popular form of treatment. The idea was that how you chew, how you swallow, and how you position your tongue, repeated thousands of times a day for your entire life, would result in changes to your teeth and your smile. Imagine if every time you swallowed you pushed your teeth out a bit; eventually your teeth should move outward. Sandra enrolled her preteen children in myofunctional therapy and marched them through the exercises. At the same time, she kept studying the literature and investigating more intensely, while keeping a close eye on the kids' development.

On a spring day in early 2012, at the recommendation of a colleague in an orthodontic myofunctional study group, she heard that one of the early founders of a practice called orthotropics, Dr. John Mew, would be giving a presentation in nearby Oakland. What she learned from Mew, the father of orthotropics, hit her with the clarity that must have first hit early scientists with the idea that Earth wasn't the center of the universe. Surely it couldn't be true, but so much indicated it *was* utterly true. Orthotropics finally explained to Sandra what she intuitively knew and what led her on her journey to find a better solution for treating her own children. While myofunction dealt with "muscle function," orthotropics dealt with posture. While myofunction was concerned with the powerful movements we did from time to time, orthotropics dealt with what we did all the time. Sandra's focus shifted more to posture, the resting state of the body, and by promoting the right oral posture she could finally address the cause and not the symptoms. So when Sandra started listing all the symptoms, Paul at first couldn't believe that something so simple could cause such an epidemic. How could poor oral posture be a linchpin to so many diseases?

After several weeks of dinner club discussion, the importance of Sandra's work became evident to Paul, as did how it fit into his long-term interest in human evolution and the environment. It even took Sandra a while to see why he was so excited about the connections among chewing, stuffy noses, and smiles. But he had spent his life connecting things such as population, food, toxins, resources, water, weather, war, and politics into a unified picture of the human future. When we finally took our idea for *Jaws* to an editor, he said, "Let me get this straight: no orthodontist is practicing this, you are the only ones who know about this, and you think that everyone needs to beware of a 'huge public health issue' right under their own noses?" Yes! The clincher for him, and many others, was Image 4, showing what a hunter-gatherer's jaws look like, with roomy perfect arches of well-aligned teeth, with no impacted wisdom teeth—a movie star's dream jaw, 15,000 years before movies!

It's important to note that the two of us had no idea a jaw-based "epidemic" was happening until Sandra discovered its symptoms in her own children. Like the vast majority of people, even with our long-term scientific interests in public health, we had no awareness of an epidemic that

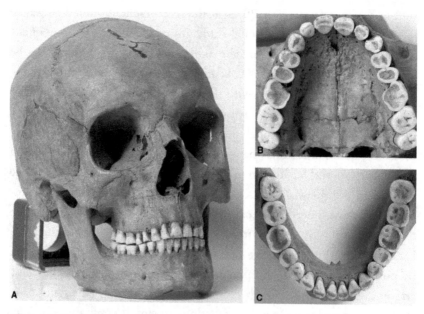

Image 4. Skull of a pre-industrial human being with a spacious jaw and all molars fully in place. People before industrial civilization did not get impacted wisdom teeth. Male skull shown is from a 14th century church in Oslo. Note lack of crowding and absence of malocclusion. A, Skull with jaws; B, upper jaw arch; C, lower jaw arch. (Photos courtesy of *American Journal of Orthodontics and Dento-facial Orthopedics*)

could, alongside issues such as obesity and type II diabetes, be of substantial importance. The "jaws" epidemic was concealed behind the commonplace. Its most obvious symptoms are oral and facial: crooked teeth (and the accompanying very common use of braces), receding jaws, a smile that shows lots of gums, mouth breathing, and interrupted breathing during sleep. A bother, but hardly an "epidemic"—at least not until one recognizes that underlying these symptoms are frequently very serious diseases, many related to the stress of poor sleep. They include heart disease,[7] eczema,[8] lowered IQ, depression, attention deficit hyperactivity disorder (ADHD), and perhaps even Alzheimer's disease.[9] One important reason for the obscurity of the epidemic has been that evidence on the frequency and strength of the connection of those diseases to oral-facial issues often is quite difficult to obtain.

Usually health scientists depend on a statistical association, rather than clear knowledge of cause-and-effect mechanisms derived from experiments. For example, one seven-year investigation was done on middle-aged men in Sweden with sleep apnea—pauses in normal breathing when sleeping. Sleep apnea occurs when breathing during sleep is interrupted and the quality of sleep is impaired (during episodes the victim often moves from deep to shallow sleep). The afflicted men were found, when other likely causative factors are eliminated, to have more heart problems than those with uninterrupted sleep. In addition, effective treatment of the sleep apnea reduced the chances

Image 5. Nobody would dispute that these two kids are drop-dead-gorgeous. Caretakers need to be trained to notice the subtle sign—their gummy smiles—that growth is not progressing in the right direction. (Photo by Gorete Ferreira.)

of cardiovascular problems.[10] A similar Swedish study strongly suggested a causative connection of sleep apnea to coronary artery disease and stroke.[11] Also suggestive, and frightening for those with sleep apnea, 46 percent of sudden deaths occurred between midnight and 6 am. In those without sleep apnea, only 21 percent of deaths occurred in that nighttime period.

The main kind of sleep apnea, obstructive sleep apnea (OSA), is due to physical blocking of the airway. It seems to be increasing and has become a significant factor in public health. Some 20 percent of American adults are afflicted, and about 3 percent have a sufficiently serious case to cause daytime sleepiness. But sleepiness is the least of it. As many as half of cardiac patients suffer from the disease.[12] Sleep apnea also appears to generate mental problems, including lowered IQ, shortened attention span, and difficulties with memory.[13] Sleep apnea is often not diagnosed, and statistics have not often been gathered on its prevalence, age of onset, and presence in the medical histories of individuals who develop other various chronic diseases that may be related. Furthermore, evidence on mechanism, on, say, how interrupted breathing during sleep might make an individual more susceptible to a disease like Alzheimer's, is frequently absent. These sorts of illness, as we'll see, are what we're concerned with in relation to poor development of the jaws, and the impacts of that developmental deficit on the face and the airway. But gathering more detailed information will be slow and difficult. Experiments won't be used. No doctor is going to interrupt the breathing of large samples of people for a long period and compare their fates to "control" individuals not subject to nightly strangulation. You may be able to guess why not! Similarly, we would not suggest subjecting children to practices that we believe cause malocclusion to test our own theories about the jaws epidemic.

Escalating attempts to straighten teeth, to treat one of the epidemic's most prominent symptoms, are one obvious indicator of the scale of the epidemic. Having braces as a child has become so common in the Western world that it can seem a rite of passage. Today an estimated 50 to 70 percent of children in the United States will wear braces sometime between the ages of 6 and 18.[14] It is not clear how much the increase in use of braces in recent years is a response to a great explosion of malocclusion or a consequence of less expensive tooth-straightening appliances, better marketing by dentists, and changes in attitudes on appearance in a photo-addicted society (think "selfies."). Ironically, the effects of braces may not always be as beneficial as people have been led to think. As we'll see, braces may actually reduce the size of the airway,[15] leading eventually to problems in breathing like sleep apnea.

That the diseases just noted are related to modern civilization is strongly indicated by the near absence of their symptoms in the evolutionary and historical record. Our hunter-gatherer ancestors had spacious jaws, with a continuous smoothly curved arch of teeth in each jaw, including third molars ("wisdom teeth") at the back ends of the arches. Indeed, Stanford evolutionist Richard Klein, a top expert on our species' fossil record, has told us that he personally had never seen an early human skull with crooked teeth. Further, the oral-facial epidemic of modern times, although rooted, we believe, in the agricultural revolution, was very slow in starting. Recently a cemetery of common people of the Amarna culture of ancient Egypt, dating to more than 3,000 years ago, was discovered. The skeletons had the tooth wear characteristic of farming peoples, the investigators noted,

Except for the occasional slight incisor crowding and rotation, observation of the teeth indicated that they were well-aligned with very-good-to excellent occlusion, in general. Thorough analysis of dental data from the Amarna burials has shown that Egyptian and most ancient teeth have extensive tooth wear on occlusal (chewing) surfaces of even the youngest individuals. Malocclusion is rare in Amarna but very common in America; tooth wear is extensive in Amarna yet rare in America.[16]

There is a common and serious misapprehension about malocclusion. As one friend said, "We take it for granted that malocclusion is genetic—we've always considered my son got his crooked teeth from my wife." As you will see, virtually all the evidence shows that the oral-facial epidemic can be traced *not* to our genes but to changes in our culture, particularly to ones in how and what we eat and where we live. These have changed greatly from those of the Stone Age, in complex patterns starting around the time people began to settle down and practice agriculture.[17] As anthropologist Clark Larsen put it: "There has been a dramatic reduction in the size of the face and jaws wherever humans have made the transition from foraging to farming."[18]

With proper attention to our children's diet, eating habits, breathing patterns, and what we term "oral posture" (how they hold their jaws when

not eating or speaking) many aspects of the epidemic could be amelio-
rated or avoided entirely. Jaws could return to their hunter-gatherer and
Amarna patterns of growth. There's much that alert parents can do for
their children and some things also that adults can do for themselves that
can help to at least reduce the likelihood of some diseases like heart attacks
and cancers.[19] An array of evidence indicates childrens' future well-being
can be greatly enhanced by encouraging a few simple habits early in life.
Consider how you yourself breathe, chew, and position your mouth when
not speaking or eating. Being aware of that can add up to better habits that
might positively change your life, improve your health, perhaps make you
more attractive and successful, and transform how you feel about yourself.
The key is first to develop awareness of those jaw-related habits that may
eventually cause dramatic changes for the worse and then understand what
can be done to kick those habits and create a better future for your family
and yourself. That is the goal of this book. In it we will lay out the evidence
indicating that an industrial lifestyle explains the epidemic of oral-facial
health problems[20] and discuss what remedies can be found. The views expressed here are not typical of the dental and orthodontic mainstream, but
we feel these somewhat heterodox ideas need to get a hearing.

There is some history for the minority view we present, especially in the
work of pioneering orthodontist John Mew, to whom Sandra took her son
after hearing his lecture in 2012. Mew successfully treats patients by returning distorted oral-facial growth to its normal course through "orthotropics,"
a program that encourages normal jaw growth and development. Orthotropics is a very important discipline with a lousy name. It is too easily confused with standard "orthodontics," from which it has major differences.
As a result Sandra renamed "orthotropics," calling it "forwardontics," to
avoid the confusion. The two names are synonymous. *Forwardontics* is the
term we will use from now on, except when we refer to Mew's work or to
literature that employs the designation orthotropics. Forwardontics is more
descriptive for the general public and includes all treatments that focus on
forward development of teeth and jaws in both children and adults.

The problems of modern jaw-face-airway development were uncovered through the work of a series of dedicated scientists and practitioners,
including Mew, observing dramatic changes over time in facial structure
and in the incidence of chronic diseases, checking them against evidence

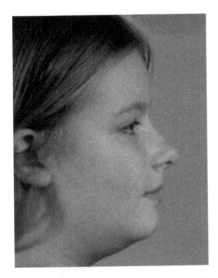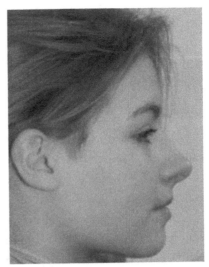

Image 6. Treatment results of orthodontist John Mew. (Courtesy of John Mew.)

in different historical periods and in different cultures, conducting animal experiments, drawing on general knowledge of human genetics and development, arriving at reasoned conclusions based on that wide array of evidence, and applying what they have learned. This has given scientists greater understanding of the oral-facial health epidemic and the changes required to end it. But in this case there has been little attempt to bring these complex results to the general public as an integrated story—something we hope to accomplish with *Jaws.*

The bottom line of our narrative is that your health and happiness, and more likely that of your children, may be at risk due to habits to which most of us never give a second thought. So here are some of the key questions you could be asking yourself:

- Are the teeth of your upper and lower jaws usually in contact or apart? Are you breathing through your nose?
- Do you usually sleep through the night?
- Does your partner complain about you snoring?
- How many times does your child chew each mouthful of food?
- Is it good to wean your baby onto special "baby foods"?
- Does your child almost always have a stuffy nose?

- Does your little girl's face make a "weird expression" when she's swallowing?
- Does she sleep with her mouth open?
- Does she make a tangle of the covers?
- Is she often tired?

How we eat can be just as important as what we eat. *How* we breathe can be just as important as what's in the air we breathe. *How* we sleep can be just as important as how long we sleep. These are all aspects of oral-facial health.

Are these dire warnings already starting to sound like diet alarms you've read before? Foods you were told to avoid yesterday, only to find that the warnings were passing fads, that "further research" reversed the advice? Fat is good for you, fat is bad for you, fat is now good for you again. Coffee is good for you, then bad, now good. Gluten is bad, vitamin E is good, and so on. As you read *Jaws*, the information and advice might sound familiarly "diet-y." But *Jaws* is not just offering faddish advice. True, some of the information in *Jaws* has been around for a long time., simple advice that may start to sound like your mother telling you to "eat with your mouth closed," "sit up straight," and "chew your food." Well, Mom was onto something; she didn't realize that what she was asking for was not just about manners and being polite; it was one dimension of forestalling a trend that has now become a public health problem. The oral-facial epidemic has developed over centu-

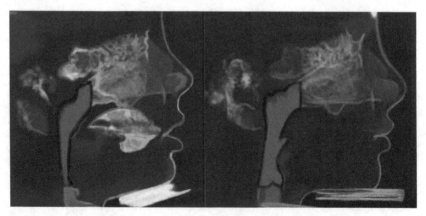

Image 7. Improvement of the airway in a boy after orthotropic and postural treatment. This is the impact mentioned in image 6.

ries, but it has accelerated as a result of common practices associated with our highly industrialized Western civilization that, since World War II, has taken over the world. So, not surprisingly, you won't find any one-step solution in *Jaws*, but rather a more sophisticated overview of a complex problem with suggestions on how to prevent and treat it—and a challenge to think.

It is often said that the face is the window to the soul, but it is also a window on the health status of the person behind the face. The human face provides visible signals that could indicate serious underlying health problems. Not only can problems in oral-facial health be an indicator of problems in the rest of your body, but they can also be a determinant of how good you look. The habits that can make faces unattractive in our culture are, sadly, the habits that can make bodies unhealthy.

Overall, then, an unheralded change has been taking place in our society. We are altering our faces with surgery and braces and other technological means when the real changes we need to make are modifications in how we ordinarily breathe and eat and sleep. They can have a more significant and lasting improvement on the quality of our looks and our health than a plastic surgeon's talents. Using quick fixes to solve our health problems and adjust our smiles may in some cases actually lead to additional problems in the long term.

Jaws starts with a chapter describing the transition from healthy Stone Age jaws to often sickly modern ones—an example of cultural evolution (change in the nongenetic information groups possess). It deals with the longstanding "nature–nurture" question, as it applies to oral-facial change. Chapter 2 focuses mostly on chewing but dips into other factors like allergies tied to the oral-facial epidemic. Chapter 3 looks at the significance of what we chew, how we chew, and where we chew (in a house or in a forest). Chapter 4 tells how attractiveness ties into jaw health. Chapter 5 discusses how and why changes occur in human jaws, faces, and oral posture. Chapter 6 brings mouth breathing and its ills to the center. Then in Chapter 7 we go to the personal level and indicate things you and your family can do to keep the epidemic at bay. In Chapter 8 we discuss how to recognize the effects of the epidemic, and we give an overview of where you might find help from dental professionals. In Chapter 9 we expand our view to ask what changes in society's culture can be made to help people like you deal with the epidemic.

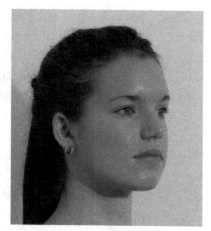

Image 8. The child on the left was told that she needed surgery to correct her receding chin. Orthotropic treatment with Biobloc and posture exercises produced the dramatic stable results in the right. Orthodontics or orthopedics is not known to produce any jaw changes remotely as impressive as these. (Courtesy of John Mew.)

Throughout the book, as scientists we naturally refer to the scientific literature describing jaw development, cultural practices, various eating/breathing environments, and related health and appearance issues. As is often the case when natural science and social science are considered together, this literature remains, sadly, rather fragmentary. In part this is because there are important ethical restrictions on making controlled observations or (especially, as suggested with sleep apnea) doing experiments on our fellow human beings. There also are logistic and, in particular, funding constraints on conducting "prospective" studies, those in which subjects are selected in advance and then followed through time. Prospective studies are the gold standard of research on health in human populations. Getting hundreds of subjects to accept the inconvenience of behaving in a certain manner (such as eating a special diet for long periods), keeping careful records, and being repeatedly interviewed, is neither easy nor cheap. Prospective studies require lots of money and patience, with periodic follow-ups that may go on for years.

Thus most studies are of the less revealing, but easier and much less costly, backward-looking type known as "retrospective"—for example, asking adults about their eating habits when they were young and then comparing the present state of health of groups who, say, were vegan or who loved steaks growing up. Such studies can yield a lot of useful information, but they have some inherent limitations. For instance, how accurate are the memories? Could the responses be biased by people reporting what they think the interviewer wants to hear? If the same questions had been worded differently, would the answers have been different?

Forwardontic (orthotropic) research, investigating the techniques used by John Mew and his colleagues, faces these problems and then some. Orthodontics is at least a clear-cut, professionalized, medical/dental treatment, engaging a large group of practitioners, and has thus been the subject of more or less standard medical research. Forwardontics is primarily a postural discipline, pursued by a small cadre of orthodontists and dentists. It is harder to practice than conventional orthodontics and less likely to be financially profitable, and its successes are highly dependent on patient cooperation. For those reasons forwardontics (as orthotropics) has been relatively ignored by the research community, and conclusions about forwardontics often need to be drawn from small samples, certain types of anecdotes, photographic histories of patients who sought help (not, then, a random sample of individuals), and the like.

Because of these limitations, some caveats should be mentioned. We are focusing on problems that seem to be related to oral-facial development in a rapidly industrialized world. Answers to some of the issues are relatively clear cut, for example that the modern environment in which children develop, especially how they chew, how they rest their mouths when not eating or talking, and the allergens to which they are exposed can greatly influence the development of their jaws, faces, and airways. It also seems likely that oral-facial responses to the relatively new industrial-era environment are largely responsible for increases in sleep apnea, which is known to be very stressful. In turn, stress notoriously contributes to an

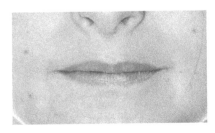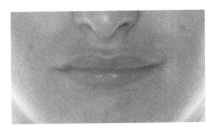

Image 9. Lips become more full when the jaw is forward and posture is corrected (12-month change).

array of serious chronic diseases. But evidence of the scale of stress and of its contributions to those diseases or the mechanisms that make the connection is often thin to nonexistent.

In some cases we've simply speculated—we suspect that snoring in hunter-gatherer children was rare, for example, but we haven't found evidence to support that notion—no reports from, say, people who lived with !Kung people on their sleeping habits, no evidence that leopards were attracted to snoring kids during humanity's long trek out of Africa. Considering the relationship of snoring to modern jaw configuration, mouth breathing, and similar factors,[21] we thought it likely that our hunter-gatherer ancestors were not heavy snorers. In any case, we've tried to make clear, at least by context, when we are simply speculating, and our reasons for doing so.

In short, *Jaws* is designed to introduce you to the vast problems of oral-facial health, which, like issues of gluten, might have once seemed as simple and insignificant as sliced bread and which now are as significant as sliced bread. And it is designed to help you decide what, if any, personal actions you can take to improve your health and well-being. It's a guide for thinkers, **not** a recipe book. So read on and make up your own mind.

CHAPTER 1

PRIMITIVE BIG MOUTHS
TO MODERN MALOCCLUSION

Jaws may remind us first and foremost of sharks, but human jaws are really at the center of our story here. Our upper jaw, which is technically known as our *maxilla*, seems as if it is just the base of our skull, but it is actually formed by two bones, one on each side, fused together. Our lower jaw, technically the *mandible*, is likewise made by the fusion of two bones. If the jaws develop correctly they have ample room for all of the teeth, and the teeth fit together well. Both upper and lower jaws can move and change in the process of development. That process has been gradually altered ever since our ancestors began to use tools, cook, cease their mobile hunting-gathering lives and settled down to practice agriculture some 10,000 years ago; they then marched on to create the civilization we know today. The superficial result, as we have seen, is malocclusion.

The evidence shows that some common notions about malocclusion are wrong. A predominant story is that malocclusion is caused by bad mixes of genes. It goes like this: as people moved around Earth after leaving Africa tens of thousands of years ago, groups with different characteristics mixed, and matings between men with big teeth and women with small jaws produced kids with malocclusion. But in fact badly fitting teeth are *not* usually caused by bad genetics or by parents having contrasting genetic endowments for facial structure—say, inheriting dad's giant teeth and mom's dainty jaws.[1] Dr. Hal A. Huggins put the "mixed-genes" argument in context in his book *Why Raise Ugly Kids?*:[2] "Horse and donkey—cross them and you get a fine work animal. Used them a lot on the farm and know what? I never saw a mule with horses' teeth and a donkeys' jaw."

With extremely rare exceptions everyone is born with the DNA that allows normal development of teeth, jaws, and tongue. After all, for millions of generations individuals with a working combination had more offspring than those who couldn't eat so well—natural selection in action. The DNA of those successfully reproducing ancestors, inherited in equal measure from each parent, contained the genes that enable the assembly of all the standard parts of an adult human being. Through that very long process of natural selection—people with some DNA endowments having more kids than those with other endowments—kept children's development occurring in ways that produce a harmonious whole in a wide variety of natural environments. That's why there's a pronounced shortage of children with one blue and one brown eye when blue-eyed and brown-eyed parents mate, and why few offspring of football linemen mated with petite women have massive shoulders and spindly legs.

DNA function requires the availability of appropriate molecules with which to create cells, tissues, and organs. The assembly of a healthy person, a good reproducer, depends also on a safe and supporting womb in which the first nine months of development will occur. And natural selection has led our species to function in a postnatal environment in which the developing person will gain nutrients by eating in a certain manner. The individual will learn to crawl, toddle, and walk, providing the environment that will interact with the DNA to produce strong leg muscles. The person will also, in the right environment, counter the force of gravity tending to drag down the lower jaw. That "right" environment, we argue, is created by tough

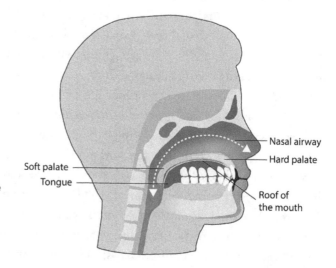

Image 10. Proper oral posture: tongue resting on the palate, the lips sealed and the teeth in light contact between four and eight hours a day.

Nasal airway
Hard palate
Soft palate
Tongue
Roof of the mouth

foods, much hard chewing, and, when not speaking or eating, holding the mouth closed with teeth in light contact and the tongue resting on the roof of the mouth (palate). That's the environment in which the jaws should spend most of their time but especially at night when growth takes place.

Genetic endowments take many generations to evolve in response to novel environments, and they only do so when individuals with some DNA configurations reproduce much more than those with others. In other words, members of industrial societies—us—still must work with DNA plans evolved for building individuals in a hunter-gatherer environment. We have entered the space age with Stone Age genes that evolved to produce jaws adapted to a hunter-gatherer diet. This has had some unfortunate consequences.

In the Stone Age eating and jaw-resting environment that lasted for 3 million years or so, human DNA evolved into plans for wide upper and lower jaws in each of which all the teeth fit without crowding and the jaws met without malocclusion. That DNA, interacting with that environment, resulted in ample airways. Since the agricultural and then the industrial revolutions, however, that eating-resting environmental pattern has been dramatically altered. Societies have culturally adapted to changes like the easy availability of softer weaning foods after the invention of agriculture and the comfort and safety of moving "indoors" once perpetual motion in search of food was no longer required.

Does this mean that "environment" is more important than "genes"? Not really. To simplify thinking about gene–environment interactions, one can imagine that a person is like the area of a rectangle—a product of the width (genetic plans) and the length (environment in which those plans are executed). One cannot say that either the width or the length of the rectangle is "more important" in creating the area any more than one can say that DNA or the environment is "more important" in creating little Hendrik. The area of a football field can be doubled by doubling either its length or its width. If either happens we can say what caused the change, but that does not alter the fact that both width and length determine the area.

If Hendrik's mother almost died of starvation while she was pregnant during the famine in Holland at the end of World War II, we can say his comparatively low birth weight was caused by a change in the environment. We can't say that his weight was caused more by his environment than by his genes. With food scarce, the interaction of genes and environment changed. That's why we can assert that our species has brought genetic plans for Stone Age jaws into the 21st century—the jaws are a product of genes and environment, but the environment has changed dramatically over the course of a few millennia while the genes have not. The result is a decline in what we've called *oral-facial health*, and this is why we've had to look elsewhere than major genetic change to explain the rise of malocclusions and other modern oral-facial problems.

As you read *Jaws*, it would be good to keep all this in mind and to remember that there is a great deal of variation from person to person, rooted in genetic and cultural differences, personality, divergence of personal experiences, and how these and other differences interact. Not everyone has poor oral posture, not everyone with poor oral posture will suffer serious consequences from it,[3] and not everyone with oral posture problems will be able to solve them.

As far as we can discover, the changes in jaws with new diets and urbanization were first noticed and recorded in the 1830s. For two decades before the American Civil War a Philadelphia attorney named George Catlin, a talented artist, made a series of trips to the American West that would make him famous as a painter and ethnographer (describer of the culture) of Native Americans. He saw a group of Native Americans passing through Philadelphia, became fascinated by them, and decided to document their ways of life. He ended up claiming to have visited 150 tribes with more than 2 million members throughout the Western Hemisphere. Catlin's portraits of Native Americans, made before their cultures were altered by contact with European cultures, form an invaluable archive, now housed in the Smithsonian American Art Museum.

In his travels among those who had been relatively isolated from the culture of the European settlers he was struck by the difference in facial structure and bearing of the Native Americans compared to the people of European background he had grown up with. He preceded modern scientists like Richard Klein in noticing changes in the jaws of preserved skulls. Among the Mandan Indians, a tribe of 9,000 individuals, he examined several hundreds of bleached skulls. "I was forcibly struck with the almost incredibly small proportion of crania of children; and even more so, in the almost unexceptional completeness and soundness (and total absence of

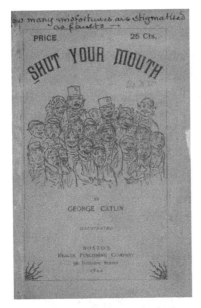 

Image 11. *Shut Your Mouth and Save Your Life*, written and illustrated by Catlin in 1861. Catlin's title indicates how even in the 1800s he understood how important closing the mouth is for health. George Catlin (1796–1872) self-portrait (right).

malformation) of their beautiful sets of teeth, of all ages, which are scrupulously kept together, by the lower jaws being attached to the other bones of the head." [4]

He noticed the Native Americans slept outside, and they kept their lips closed nearly all the time. Their women breastfed, and as soon as her baby was off the nipple after feeding, the mother would close the infant's lips with her fingers. This was not something the mothers of European background did. The natives called the whites not only "palefaces" but also "black-mouths" because their mouths were so often open, their lower jaws hanging down. Further, those Indians not in close contact with European civilization were, Catlin noticed, typically much healthier than the newcomers to the continent. He saw the scarcity of children's skulls in burial collections, and interviewed many older people on childhood mortality. From Mandans,

I learned from the Chiefs, that the death of a child under the age of 10 years was a very unusual occurrence; and from an examination of the dead bodies in their Cemetery, at the back of their Village, which were enveloped in skins, and resting separately on little scaffolds of poles erected on the prairies, amongst some 150 of such, I could discover but the embalmments of eleven children, which strongly corroborated in my mind the statements made to me by the Chiefs, as to the unfrequency of the deaths of children under the age above-mentioned; and which I found still further, if not more strongly, corroborated in the collection of human skulls preserved and lying on the ground underneath the scaffolds . . . The instances which I have thus far stated, as rather extraordinary cases of the healthfulness of their children, in the above tribes, are nevertheless not far different from many others which I have recorded in the numerous tribes which I have visited; and the apparently singular exemption of the Mandans, which I have mentioned, from mental and physical deformities, is by no means peculiar to that tribe; but, almost without exception, is applicable to all the tribes of the American Continent, where they are living in their primitive condition, and according to their original modes.[5]

In those days, death rates of European children were high; mortality tables in Europe during the 1850s show that about a quarter of children died by the age of 5,[6] and only one in four people survived beyond 25 years of age.[7] Mortality rates in big European cities were higher, and one assumes that in the cities of eastern North American many children also died young. Native American kids seemed to thrive by comparison.

Catlin, originally a mouth breather, observed that the Indians never mouth-breathed and were extremely healthy. As a result he changed his pattern of respiration and taught himself to nose breathe:

Who, like myself, has suffered from boyhood to middle age, everything but death from this enervating and unnatural habit, and then, by a determined and uncompromising effort, has thrown it off, and gained, as it were, a new lease of life and the enjoyment of rest—which have lasted him to an advanced age through all exposures and privations, without admitting the mischief of its consequences?[8]

He wanted to convince others of the health advantages of the Indian ways. To this end he wrote a short book, *The Breath of Life* (1861),[9] later retitled as *Shut Your Mouth and Save Your Life*,[10] which condemned mouth breathing and assigned an array of ills to it, including "derangement" of the teeth, that seems remarkably prescient of the jaw derangement that is epidemic today and creates a vast business for orthodontists. Here is another brief extract from Catlin's book that gives a feel for how serious he considered the problem to be:

And the most abominable, disgusting, and dangerous habit belonging to the human race . . . of sleeping with the mouth open, has but one certain and efficient remedy, which is in infancy . . . In advanced life, with the muscles unnaturally elongated by long and constant distention, the dislocation of the jaw is further from remedy, and the malady more difficult to cure; but even then it is possible. Bandages may be applied, and the jaw may be strapped up during sleep; but these don't shut the mouth, nor will any mechanical application that ever can be invented, do it. Temporary benefits and partial relief may be obtained in this way—yet I believe there is but one effectual remedy for the adult habit, which is, adult consciousness, and constant adult conviction that premature death is close at hand for him whose mouth and lungs, during his sleep, are open receptacles for all the malaria (and changes of temperature) of the atmosphere that may beset and encompass him."[11]

Catlin at first may appear rather kooky in some of his views, or at least "eccentric." He was writing over 150 years ago and believed malaria was related to mouth breathing and "poisonous particles in the air," a belief

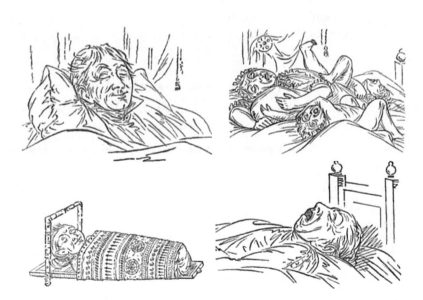

Image 12. Illustrations from Catlin's *Shut Your Mouth and Save Your Life* contrasting natural sleep (left) and unnatural sleep (right).

he shared with members of the medical profession of his day. He was also kooky for his day in his admiration and empathy for the Native Americans, whom he saw as examples of the "natural men" living in harmony with nature, idealized in the Enlightenment. He also saw them, rightly, as being abused.[12] He was, above all, a brilliant observer, way ahead of his time in recognizing many threats posed by mouth breathing.[13] And he was on to something, later research would show, even if not always for the right reasons. Anthropologists have reported that the size of the human mouth has long been shrinking.[14] Because human beings have been using stone tools for at least 3.3 million years,[15] that may represent the time during which the shrinkage has occurred. Stone tools permitted a greater shift to carnivory because the ability to cut meat into small pieces reduced the amount of chewing required to extract nourishment. Less chewing reduced the need for large, powerful jaws. The same can be said for being able to use stone tools as pestles for grinding food into small, more easily digested fragments. Cooking also reduced the chewing time needed to acquire the nourishment required to support energy-demanding big brains, but tools preceded cooking by as much as a million and a half years.

As we said at the start, Stanford evolutionist Richard Klein, a top expert on our species' fossil record, has told us that he personally had never seen an early human skull with malocclusion.[16] This pattern has been confirmed by Harvard evolutionary biologist Daniel Lieberman, who writes in his fine book *The Human Body*:

The museum I work in has thousands of ancient skulls from all over the world. Most of the skulls from the last few hundred years are a dentist's nightmare: they are filled with cavities and infections, the teeth are crowded into the jaw, and about one-quarter of them have impacted teeth. The skulls of preindustrial farmers are also riddled with cavities and painful-looking abscesses, but less than 5 percent of them have impacted wisdom teeth. In contrast, most of the hunter-gatherers had nearly perfect dental health. Apparently, orthodontists and dentists were rarely necessary in the Stone Age.[17]

Nevertheless, dental crowding has been documented in one sample of remains from France dating to a couple of thousand years ago.[18] And it was recently reported in a single early anatomically modern human from Qafzeh cave in Israel, dated to about 100,000 years ago.[19]

Image 13. Daniel Lieberman, Harvard biologist, an expert on the evolution of the human head. (Photo by Jim Harrison.) On the right is the 1000-year-old skull of a 35-year-old Philistine woman excavated in Israel, showing again how pre-industrial people lacked malocclusion. (Photo by Jim Hollander/EPA.)

These examples show that malocclusion could occur even in our distant ancestors, which is hardly surprising considering the varied environments to which they were exposed. Perhaps the early French population had an unusually soft diet; that is suggested by the relative lack of wear on the teeth. Malocclusion can also occur in modern traditional societies and has been documented in one highly inbred Amazonian population,[20] showing that proper oral-facial development can sometimes be disrupted at a population level by genetic factors. Indeed the Qafzeh individual's crooked teeth could have been related to his or her genes if there had been a high level of inbreeding in that population.

But overwhelming evidence indicates crooked teeth were extremely rare among hunter-gatherers and dental crowding less common among early agriculturalists and people of the medieval period than in industrial populations.[21] A comparison of 146 medieval skulls from abandoned Norwegian graveyards with modern skulls indicated "a significant increase in both the prevalence and the severity of malocclusions during the last 400 to 700 years in Oslo, Norway."[22] The skulls of people scored as in "great" or "obvious" need of orthodontic treatment made up 36 percent of the medieval sample and 65 percent of the modern sample. In Sweden, where 10 percent of the modern population was judged to be "in very great need" of orthodontic treatment, over 100 skulls from the Middle Ages were examined, and instances of malocclusion were "very less common" in those skulls than in modern Scandinavian people.[23]

On the jaw size issue, Swedish Orthodontist Lennard Lysell made extremely careful measurements of skeletal remains from a medieval graveyard uncovered in excavations connected with the construction of an airfield in 1951.[24] He worked with the skulls of some 97 adults representing the individuals with the most teeth preserved out of roughly 250 skeletons from the 11th to the 13th centuries. Lysell also considered whether the sample of medieval Swedish skulls was representative of the general population then and compared his skull measurements with those previously published of modern Danish and Swedish samples of skulls. His results, like those of some other Swedish investigators, suggested that there had been a detectable reduction in jaw width since the medieval period.

His basic results were confirmed in a study by dentist Christopher Lavelle comparing 210 lower jaws from skulls of the Romano-British period (43–400 CE), the Anglo-Saxon period (410–1066), and the 19th century. The size of British jaws was also declining as the modern age came into being (and the coarseness of the diet was reduced).[25] Well-preserved skulls from four or five centuries ago show almost no malocclusion. In addition, there is much evidence supporting the anthropologists' conclusion that jaws and faces do not grow to the same size and shape now that they once did.[26]

As Robert Corruccini, a leading dental anthropologist, put it, there is every indication that "increases in malocclusion have accelerated during the last 150 years in technologically advanced communities, after having shown relatively modest changes for 6000 years."[27] For instance, a comparison of the skulls of Austrian men from the 1880s and 1990s showed significantly more malocclusion in the 20th-century skulls.[28]

Much more information on diets and jaws is needed to accurately map out the initial course of what has become a pandemic of overcrowded jaws. Yet evidence that we'll discuss later shows that people switching from traditional to industrialized diets can manifest oral-facial changes in a single generation. Thus it seems likely that the progressive shrinking of jaws suddenly accelerated with industrialization. More information on the pattern of diet softening over the last thousand years would certainly be desirable to help pin down diet's contribution to jaw development. Sadly, most of the literature on dietary change focuses on the nutritional content of foods and diseases like diabetes and obesity, and not on toughness-softness and jaw development.[29] Be that as it may, the speed of the transition to the age of braces indicates that cultural rather than genetic changes have been primarily responsible for increasing malocclusion.[30]

One bottom line in this array of evidence is, as anthropologist Peter Lucas and his colleagues concluded, that changes in the toughness of diet in mammals can result in small jaws and malocclusion: "Dental crowding in modern humans is considered the combined result of tool use to comminute [pulverize] foods and cooking to modify their mechanical properties, such as toughness."[31] As we'll see, mouth breathing, especially from moving indoors[32] and increasing allergies and stuffy noses (often from circulating colds in child-care centers) in children, appears to complete the story. These cultural changes, especially the trend toward eating softer foods,[33] appear to have led in turn to progressively more alteration of jaw development[34] and in some cases too little room for the last molars (wisdom teeth) to erupt (emerge from the gum), a phenomenon known as "impacting" of wisdom teeth. Impaction of these teeth too often results in their routine and often unnecessary extraction in the United States at high cost, contributing to substantial dentist-caused disease in the form of pain, swelling, bruising, infection, and general discomfort. In addition some 11,000 patients annually suffer permanent numbness of the lip, tongue, and cheek tracing to injury to nerves during the surgery. Dentist and public health expert Jay W. Friedman estimated that roughly two-thirds of the extractions are unnecessary, "constituting a silent epidemic of iatrogenic [physician-caused] injury that afflicts tens of thousands of people with lifelong discomfort and disability."[35]

CHAPTER 2

MOSTLY CHEWING

Language with syntax made possible many human cultural developments, including the evolution of agriculture, which appeared in different parts of the world about 10,000 to 6,000 years ago and wrought a huge change in human diets. Grunts and gestures were simply inadequate to communicate concepts such as "help me dig this ditch to bring water to the seeds I planted over there." People first began to supplement the food they got from men hunting and from women gathering plant materials (especially tubers), by encouraging valued plants to grow near their camps (agriculture) and encouraging animals to hang around (domestication). This had a series of important impacts. Agriculture made it possible for groups to become more sedentary and to develop surpluses. That in turn made it possible for some to specialize in non–food-producing activities—making tools (manufacturing), guarding camp and keeping order (soldiers), educating the young (teachers), placating evil spirits (priests)—building the foundations of what we think of as civilization. Agriculture freed part of the population from the critical activity of acquiring food, and then the invention of writing ended the dependence on the human brain for information storage; together these opened the door to the evolution of modern societies. That's how we could end up with industrial diets, indoor living, books about eating, orthodontists, and authors hunched over computers.

Anthropologists and archaeologists have documented the importance of diet diversity in human history[1] and have shown that the advent of agriculture produced both dietary changes[2] and new styles of eating. For example, Dr. Loren Cordain and his colleagues discuss in detail some of the nutritional differences of foods typical of hunter-gatherer and early agriculturalist diets and correctly point out that in most cases there has been too little time for the human population to respond genetically in any significant measure to resulting new selection pressures (a prominent exception has been the evolution of adult lactose tolerance in dairying populations[3]). Changes have clearly occurred since our ancestors settled down in such dietary elements as glycemic load (related to carbohydrate content), availability of different nutrients, fiber content, and food processing. Cordain is noted for his promotion of the controversial Paleolithic diet, featuring primarily higher levels of protein and lower levels of carbohydrate, which has shown some benefits in studies of people following "Paleo" diets.[4] Interestingly, as we have noted, this rather extensive dietary literature remains largely silent on the history of diet toughness and required chewing, even though it is clear that it has an important impact on oral-facial development and led to the shrinkage of jaws.[5]

After people settled down, no longer was the main eating tool a knife, used to slice off a chunk of meat held between the hand and clenched teeth.[6] Spoons doubtless appeared first because they were simply modifications of natural artifacts: shells, usefully shaped pieces of wood, and so on. Forks appeared later, probably as devices for manipulating meat while it was being cooked. Both spoons and forks were around in ancient Egypt and China, and chopsticks trace back to the Neolithic in China.[7] Overall, utensils are associated with softer and more fragmentary diets. Instead of mostly chewing tough meat, people began to consume cooked rice and other foods and to reduce tougher food to small pieces on their plate. Fruits evolved, under artificial selection by farmers, into softer, sweeter food items.[8] Hard chewing, which took up something like half the day of our chimplike forebears who hadn't discovered cooking, certainly declined with the use of fire to prepare food. It became less and less frequent in farming communities and then in industrialized societies.[9] As evolutionist Daniel Lieberman summarized the results: "The mechanical forces generated by chewing food not only help your jaws grow to the right size and shape, they also help your teeth fit properly within the jaw."[10] Changes in chewing changed human jaws and faces.

Genetic evolution is slow, and the need for a tough "chewing environment" to properly express the genes for oral-facial development persisted even as cultural evolution reduced the need for prolonged chewing to obtain nourishment. Exact chronologies of the softness of diets as the West industrialized have eluded us, but we do have clues. For instance, human beings had long had a sweet tooth, but early after the agricultural revolution it could be satisfied mainly by raiding beehives—a limited resource that had to be acquired over the objections of the bees. The rich among the ancient Romans used honey extensively in their very varied diet. Poor Romans had much simpler diets, featuring especially bread and thick stews.[11] In the Middle Ages honey was used, especially by the wealthy, in a great variety of cakes, custards, tarts, fritters, and the like—soft foods all.[12] Sadly we have not found any accounts of rates of malocclusion among rich and poor in medieval or Roman times. Honey was a rich man's sweetener until the appearance of sugarcane and especially until the sixteenth-century European occupation of the Caribbean. That area was ideal for cane production, and with the horrific slave trade supplying the labor, "white gold" became a major factor in commerce; prices dropped until common people could have abundant soft, sweet foods in Europe.

We speculate that the big jump in dining on soft foods (and malocclusion) started in the 19th century. That's when the meat grinder was invented and hamburgers became a staple, ice cream was first wildly popular,[13] mass-produced baby foods were first marketed,[14] and, not coincidentally, we guess, canned foods became more popular.

Lifelong consumption of an industrialized (softer) food supply extending into adolescence and adulthood further exacerbated the problem of developing smaller jaws with crowded and ill-fitting teeth. Some people think there is lesson for us in a famous 1930s experiment by Dr. Frances Pottenger whose subjects were not people but cats.[15] Pottenger fed one group of cats soft cooked food and pasteurized milk and compared their development to that of cats that ate their traditional raw meat diet. The cats reared on cooked food grew to be smaller than their meat-eating counterparts, developed health problems, and were unable to reproduce.[16] Unhappily, there were many unavoidable flaws in this 1930s work. Pottenger lacked information on feline nutrition, and more recent experience shows that cats thrive and are prolific on cooked cat food today.[17] And, of course, people are very different animals from cats!

Some dentists followed Catlin's pioneering path of comparing the oral-facial health of indigenous peoples with that of those eating a Western diet. In the 1930s, the founder of the research section of the American Dental

Association and the association's chairman from 1914 to 1928, Weston Price, went around the world looking at the teeth of people in native communities and found that members of traditional societies typically did not suffer the widespread tooth decay, gum disease, or crowding of teeth common in America. His approach was important, but many of his results on topics like the incidence of dental decay and the relationships to modern diets have remained controversial.[18] Price's theory was that tooth decay, gum disease, and the like, escalated with the adoption of a Western diet with refined flour, sugar, and pasteurized milk,[19] and this has elements of truth, as we will see. But his main contribution was to compare people in traditional societies with those in urbanized industrialized societies. He noticed a lack of malocclusion in the former; even the wisdom teeth, so often troublesome in the West, had room to fit comfortably in the wide smooth arch of teeth in a normal indigenous mouth.

He was wrong, however, in the cause of those differences, which he assigned to the nutritional composition of the different diets. He noted that shape of indigenous peoples' faces changed in as little as one generation with a shift to Western diets, but missed that the central dietary issue related to jaw structure is not which nutrients it contains, but how much chewing it required. This is illustrated by the photo of two brothers who moved to an industrialized area (Image 14). As Daniel Lieberman put it:

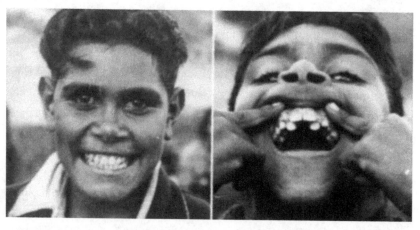

Image 14. These are two brothers who moved from their traditional environment to a reservation, where they ate a more industrialized diet. The younger, at right, shows significant dental malformation because he moved at a younger age. (Photos from Weston Price.)

"For millions of years, humans had no problem erupting their wisdom teeth, but innovations in food preparation techniques have messed up the age-old system in which genes and mechanical loads from chewing interact to enable teeth and jaws to grow together properly."[20]

Today we can follow in the footsteps of the pioneering scientists by observing the development of malocclusion in families that move from traditional communities to industrialized ones. Image 15 compares a grandfather who grew up in a traditional community in India with a son and grandson who moved to the West to London. Note that the grandfather, who ate a traditional diet (and presumably was breast fed for a long period and weaned onto adult foods that required chewing) had well-developed jaws that sat forward in the face, whereas the son and grandson, who presumably were eating a more Westernized diet, have jaws that are set back from the position shown in the grandfather. As a result, they may have had a higher probability of developing health issues with their airways and the sleep problems characteristic of Western societies.

Corruccini expanded on Weston Price's conclusion that "the consumption of a Western Diet or 'Industrial Diet' may be one of the factors responsible for the swelling epidemic of dental problems in the human populations with western diets."[21] In the industrial world, with the development of modern medicine and sanitation around the early 20th century, infectious

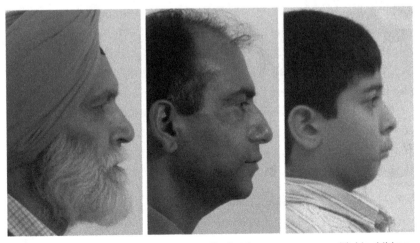

Image 15. A grandfather who had come to England as a young man with his children. His grandchild was born in an industrialized society. You can see a progressive reduction in the forward dentofacial growth in the three generations.

diseases tended to be replaced as public health problems by chronic ailments, a shift known as the epidemiological transition. Corruccini documented a concurrent transition from predominantly normal occlusion to malocclusion as populations industrialized .[22] He did research all over the world documenting aspects of this shift. In one of his studies, for example, he reviewed genetically similar populations in India, one rural and another urban, and found that those eating more refined foods had smaller jaws and had more problems with their teeth than those with more traditional diets with tougher foods.[23]

His work cast doubt on the Begg hypothesis,[24] popular in the 1950s, that malocclusion was the result of the lack of grit in the modern diet so it did not wear teeth sufficiently, they therefore grew too large for jaws.[25] One of Corruccini's early papers compared groups of squirrel monkeys (considered an excellent primate to serve as a model for jaw development in human beings) on nutritionally equivalent soft and hard diets. The idea was to test for differences in jaw development caused by how much chewing was required. He found that the monkeys on soft diets, with less exercise for their jaw muscles, suffered malocclusion. They had "more rotated and displaced teeth, crowded premolars, and absolutely and relatively narrower dental arches." At the same time there was no difference in wear on the sides of the teeth between those on soft and those on hard diets, as would have been expected under the Begg hypothesis.[26] Dental anthropologists Jerome Rose and Richard Roblee[27] confirmed Corruccini's conclusions in human beings. "Most modern malocclusions are caused by disparity between jaw size and tooth-arch length (total room required for a complete set of teeth)." they found. Such malocclusions, you'll recall, were rare in Amarna and among ancient people worldwide.

Corruccini's more recent work on human beings also confirmed his conclusion that chewing was what counted in jaw development. For example, in a study of a group of Australian Aborigines in the first generation to eat a soft, nonwearing diet, he showed:

Longer (unworn) teeth did not relate to crowding in general nor to crowding in relevant local areas or during developmental stages. Unfavorable leeway space (that remaining space after the baby teeth drop out) did not relate clearly to crowding or other malocclusions. Lowered correlations among structures and narrowness of the maxilla (upper jaw) related more significantly to malocclusion. These results are in keeping with recent thinking that small jaws rather than large teeth underlie tooth/arch discrepancy.[28]

There is now substantial evidence that another aspect of the epidemiological transition may also have an effect on oral health. The communities of bacteria that lived in our mouths for the millennia of hunting and gathering were critical to avoiding tooth decay. These bacterial communities were significantly transformed by the change in their environment when relatively sedentary hunter-gatherer groups started collecting wild plants rich in fermentable carbohydrates, and later by human diet changes associated with farming.[29] The greater availability of carbohydrates and sugars, made more extreme by food processing, favored decay-causing bacteria, which now dominate the ecosystems of our mouths. Although the direct effects of the changing bacterial flora on the size of the jaw may be minor or even nonexistent, the "drifting" (natural movement from interactions of teeth with one another) of rotting teeth if not promptly treated will influence chewing and jaw development. A child with a painful cavity will keep the teeth out of contact to avoid pain.

Keeping the teeth out of contact will make the face grow longer. Who cares? Sadly, having a longer face will result in a more restricted airway and possibly prepare a child to suffer sleep apnea. How does this happen? Keeping the teeth apart leads to abnormal development of the jaw. That's

Image 16. When the tongue lies in its resting place, pressed against the palate, it acts as a scaffold to keep the dental arch in a "U" shape (left). If the tongue is kept too low, the dental arch will narrow, causing teeth to become crowded and constricting the airway.

because teeth that are not in contact will tend to grow further out from the gum. When contact of the opposing teeth is causing pain, the tongue tends to drop to reduce the discomfort as a shock absorber draped over the teeth. With the tongue moving away from its ideal resting place fully pressed against the palate (upper jaw), the dental arch loses its "scaffold" and becomes more crowded. This, in turn, will have the effect of elongating the face as the upper jaw moves down and back, as a result of the lack of the normal contact of its teeth with those of the lower jaw. Because the lower jaw is hinged to the upper, the upper moves the lower back as well as down. This backward motion results in restriction of both the space for the teeth and tongue and the size of the airway

You might ask, "So what?" A longer face in a child is a symptom of the possible appearance of really serious problems like sleep apnea and all of its accompanying diseases. It can make it more difficult for the child to breathe.

The obvious question here is why our distant ancestors, with their chipped or decaying teeth giving them fits, didn't get elongated faces, crooked teeth, and narrow airways? The answer apparently lies in their coarse diet that required lots of chewing. It may have been very painful, but not as painful as starving to death. Modern kids can choose a milkshake

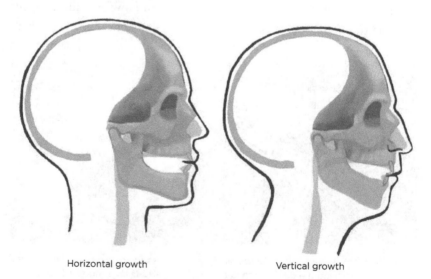

Horizontal growth   Vertical growth

Image 17. Longer-faced individuals with less defined chins have more restricted airways and are predisposed to have obstructive sleep apnea (OSA).

and avoid a tough pork chop, and they can relatively quickly get treatment that relieves the pain.

So the preponderance of evidence shows that abundant chewing in youngsters paves the way for proper oral-facial development. It is hard to recognize that such a seemingly minor, everyday activity can have such profound effects, but it can.

CHAPTER 3

THE DIET, POSTURE, AND HOUSING REVOLUTIONS

Perhaps the least-understood but most consequential aspect of human history is the great flowering of cultural evolution and development of complex language that occurred some 70,000–100,000 years ago—what Jared Diamond has called the "Great Leap Forward."[1] Suddenly (in terms of geological time) people could accomplish things rapidly not by changing their genes but by changing the body of *nongenetic* information their groups possessed—their culture. Those changes could be passed down generation to generation, by example, by word of mouth, eventually by text as well, and then photographs, TV, computers, and cell phones. Although significant genetic change in human beings usually takes thousands of generations, tens of thousands of years, significant cultural evolution can occur within a generation or two, or even less.[2] Think of how seeing the first photo of Earth from outer space changed humanity's view of its home planet. The implication of those cultural transformations is profound: more and more human beings, equipped by our deep historical and evolutionary genetic experience to live as nomadic hunter-gatherers, have taken up life in a radically different, modern industrial cultural environment.

There is no evidence that human genes involved in jaw and face development have changed significantly since the Great Leap Forward. Very few human beings suffer from *inherited* (as opposed to environmentally induced) deformities of the teeth and face. Occasionally unlucky individuals are born with a genetic deformity of the jaws. Therefore it is what we do, what our oral posture and toughness of diet were when we were young, not the genes from our parents, that largely controls the size and basic health of our jaws and related details of configuration of our faces.

In terms of the oral-facial health issues of this book, three core areas of cultural shift and influence appear paramount:

1: What we eat and how we eat it

2: Slack jaw and prevailing oral posture

3: Indoor living and how we breathe

1: What We Eat and *How We Eat* It

Our jaws are built for a Stone Age diet, but we're living in a Big Mac environment. Lack of routine chewing is exacerbated now by patterns of bolting down "fast foods" that make up a substantial portion of the diet of children in much of the developed world today. Additionally, soft foods such as fruit, yogurt, applesauce, and peanut butter do not provide the chewing opportunity that children need. The same lack of opportunity is suffered by children of parents who cut meat and other chewable food into tiny pieces because they are afraid their kids might choke. Therefore children miss out on the learning experience of how the tongue and muscles coordinate to maximize extraction of nutrients from a meal and do not get the muscle exercise that is needed for the healthy development of the jaws.

The changes in what we humans eat over the roughly ten millennia since we were predominantly hunter-gatherers have been pretty well established by anthropologists, even though the details of timing have not yet been summarized. Moving to an "industrial diet" meant going from tough meats, tubers, nuts, and fruits through a long transition to many soft foods like hamburgers, stews, soups, bread and other bakery goods, and fruits selected for sweetness and juiciness. Less well recognized are important shifts that have taken place in *how* foods are eaten and how babies are nourished. Among these changes, one seems of particular importance in explaining modern jaw-related problems common in industrial society. If you eat hard foods, curiously enough you've got to do a lot more chewing. The more you chew, especially in the childhood years, the stronger your jaw muscles become and the larger your jaws will grow. But in your high chair or in your favorite restaurant today, you hardly have to chew at all.[3] Relatively little chewing early in life can alter the development of your entire face, jaws, and airways.[4]

You don't have to be a scientist to intuitively understand the trend away from chewing; even the writers and animators of the dystopian Pixar movie *WALL•E* understood this. They showed how all foods being liquefied, hamburgers manufactured to be sucked through straws, and lack of physical activity, not genetics, changed the faces and bodies of the humans traveling through space.

What we chew, we emphasize, is closely linked to *how much we chew.* Chewing too little, as we have seen, is a major cause of the escalating problems of malocclusion.[5] The increased frequency of crooked teeth is a signal of how serious oral-facial health problems have become in society. Native populations that move from their traditional communities to modern urbanized societies[6] may develop crooked teeth within a generation, studies show.[7] In Nigeria and India, shifts to softer diets in cities were deemed responsible for smaller jaws in urban populations as compared with rural ones.[8] Animal studies show similar differences between hard and soft diets in how the jaws develop.[9] The basic conclusion, we repeat, is clear—the tougher the food, the more you chew; the more you chew (especially when young), the more normal (and spacious) your jaws are.

Image 18. Feeding soft foods is a convenient way to nourish a child but may lead to problems with the development of the jaws and airway because it bypasses chewing.

We all want our kids to be healthy, attractive, and successful, and a proper relationship to food can greatly enhance their chances of ending up that way. Part of our relationship to food is to *what* we eat, part is to *how* we eat it, and strangely enough, *part is related to what we do with our jaws when we're not eating.* We read virtually nothing about the latter. The media saturates us instead with news about potential problems in what we eat. Every day there seem to be confusing new stories about how much saturated fat it's safe to consume or whether too much sugar causes heart disease, while reports abound on the obesity epidemic and type II diabetes, warnings about poisons in foodstuffs shipped from China, hamburgers or salad greens recalled because of *E. coli* contamination, ice cream full of sometimes deadly *Listeria* germs, arsenic in wines from California and in rice in children's diets,[10] high doses of mercury in tuna, and oversized bottles of sugary sodas, to say nothing of ads for vitamin supplements and heartburn and constipation cures.

With all this attention to the healthiness of what foods we and our children eat (or avoid eating) missing is discussion of whether our children are eating them in the right way—chewing and swallowing properly to aid their development. And, as we said, little or no consideration is given to the health importance of what children do with their mouths when they are at rest.

2: Slack jaw and Prevailing Oral Posture

Breathing and sleeping problems are only lightly tied to the many vexed questions about *what* foods our kids should eat but tightly tied to how *tough* those foods are. The evidence indicates that those problems, as we have seen, are related to how much and how hard they should chew. And, counterintuitively, breathing problems also seem to be related to how children *rest* their mouths and faces. To reiterate, it's not just the food you eat and how you eat it but also what your mouth is doing when you are *not* eating, when your mouth is at rest; that's also crucial. The correct oral posture, the one that appears most conducive to jaw development, is (when not speaking or eating) holding the mouth closed with teeth in light contact and the tongue resting on the roof of the mouth (palate), as shown in Chapter 1, Image 10. The key point here is that all these habits are cumula-

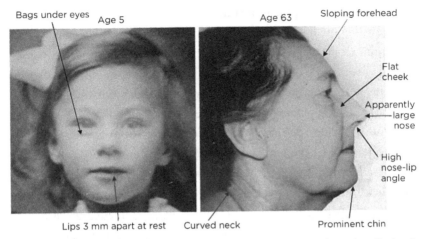

Image 19. At 5 years old, this child kept her mouth open; 58 years later the effects of bad rest oral posture can be observed. Her head had to be tilted back to keep the airway open, while maintaining the eyesight plane parallel, resulting in all the harmful anatomical and most likely physiological changes. (Courtesy of Dr. John Mew.)

tive in their effects and generally correlated. So as you chew less hard foods, your jaw muscles get slack, and thus you hang your mouth open due to an underexercised and weakened jaw muscle.

A child's oral posture determines the direction of the growth of his face and its ultimate shape and attractiveness. A face that has small jaws indicates a likely restricted airway. Health professionals recognize that a patient who has a restricted airway will tilt her head back to open the airway. The small chin is an aspect of appearance, and the inadequate airway is an aspect of health.[11]

3: Indoor Living and How We *Breathe*

As you read this, are your upper and lower teeth in light contact, your tongue against the roof of your mouth, and your lips sealed? That is, do you have what we will be calling *proper oral posture*? If not, you are probably mouth breathing, letting your lower jaw hang loose. When your mouth is open and especially if your nose is stuffed because of allergies, it is easier to breathe through your mouth. Then your jaw hangs even lower to make room for more air, and the problem compounds. Proper oral posture, which entails habitually breathing through the nose, is extremely important for the healthy development of your jaw, but mouth breathing has become common—as you'll recall over 50 percent of children can be afflicted.

Sandra recalls when she was first alerted to the issue. She accompanied her kids, Ilan and Ariela, to a theme park in Orlando, Florida. While Sandra's kids were busy running from ride to ride, Sandra took out her iPhone and started taking pictures of almost every child that walked by, each with their mouths wide open, *mouth breathing*. She grabbed her husband, her kids—she would have grabbed the stranger standing next to her if she could—and said, "Look at them, look at them all, just walking around with their mouths wide open, mouth breathing." It was true, her husband David couldn't help but notice, "It looks like a zombie nation of mouth breathers." When you are next at the mall, at the game, at any large gathering, look around you and notice the mouth breathing and look at the faces, chins, teeth that go with it, and dark circles around the eyes. It doesn't have to be this way. Look around you; how many people do you see with their mouth open, breathing through their mouth rather than their nose?

Breathing through your mouth effectively changes your mouth from a chewing device to a breathing device. A chewing device can be illustrated by two hands clapping solidly, but a breathing device requires you to curve your hands to create space for air to flow through a tube. In mouth breathers the maxilla (and hard palate) becomes narrower and develops a deep cleft—more tubelike. That is a simple way of describing how, after many millions of breaths with your mouth open, your face has changed shape.

Of course it's often tough to avoid mouth breathing, especially since we started living much of our lives indoors. Since the agricultural revolution, people, freed of the need to move frequently in search of food, began to construct more and more elaborate buildings to live in. Enclosed spaces are areas where allergens (substances that cause allergies) tend to concentrate.[12] Lots of organisms that generate allergens moved into buildings with people, such as furry pets, cockroaches,[13] molds,[14] dust mites, and a variety

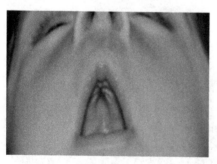

Image 20. One of the signs of airway trouble that parents should be alert to is a narrowing and deepening of roof of the mouth. (Courtesy of Dr. John Mew.)

of microrganisms.[15] In turn these allergy-friendly environments increased the odds that children would early on develop upper respiratory problems[16]—today in the United States some 60 million people suffer from them.[17] Enclosed space, like classrooms, can also increase the transmission of viruses that lead to the nose blockage of common colds. In addition, some 10 percent of children in the United States suffer childhood asthma, which can prolong periods of stuffy noses related to viral upper respiratory infections.[18] And the natural, sometimes necessary response to the stuffy nose caused by allergies is to breathe through the mouth. The same can be said, as many parents know, for the stuffy noses related to the cold viruses circulating in child care facilities. Form meets function, and the compounding effect is that over time you develop smaller jaws. The change was not a result of modified genes but of the great leap in our species' environment by moving indoors, increasing the odds that your breath will largely bypass your nasal environment.

As we'll explore, an upsurge in mouth breathing is related to an increase in malocclusion in industrial societies.[19] The faces and jaws of kids who mouth breathe develop differently from those of nose breathers,[20] which can have serious health consequences through the stress of sleep disturbance.[21] This is some of the important evidence that leads us to conclude that the way children rest their mouths, their oral posture, is a critical factor in determining their future health and appearance. Equally critical is how much and how hard they chew.

To summarize, there is a chicken and egg issue in figuring out what causes mouth breathing and related problems. Apparently some kind of allergy or other nasal blockage (inflammation) often starts early in life, restricting the nose-to-lungs airflow. This, in turn, can contribute to mouth breathing and oral posture changes such as hanging the mouth open, tilting the head back,[22] and pushing the neck forward to open the airway. Mouth breathing and hanging the mouth open together have an adverse effect on the growth and development of the jaws,[23] teeth, and face.[24] And kids who are habitual mouth breathers tend to have small jaws and may as a result develop the crowded and misaligned teeth that orthodontists struggle to correct.[25] Mouth-breathing children also may develop pulmonary hypertension, high blood pressure in the arteries that carry blood from the heart to the lungs, affecting arteries in the lungs and the right side of the heart. Pulmonary hypertension is a serious disease that can lead to death.[26]

What children eat, *how* they eat it, and how their mouths, faces, and breathing tubes develop—their oral-facial health—are aspects of children's development over which you can have great influence, especially in the critical period when children are very young. It also appears sensible to pay careful attention to allergies, especially those resulting in stuffy noses, and do whatever we can do to limit their duration. Many of the problems of oral-facial health are caused by bad habits started young.[27] Later, based on Sandra's long clinical experience as an orthodontist fixing children's oral posture, we will discuss the ways in which we may change those habits.

But, you might reasonably ask, aren't you two painting too dark a picture? After all, people in industrial societies tend to live longer, healthier lives than those in traditional agricultural societies. Bad as may be some of the trends in oral-facial health that we discuss, they clearly have not greatly reduced human well-being.[28] But we are convinced that many, if not most, people could lead happier and healthier lives if the issues we raise were dealt with, at a personal and societal level. If more attention is paid to problems with jaw development, the current generation of children may well live longer and more healthy lives than their parents. If that does not happen, the results could be grim.[29]

CHAPTER 4

APPEARANCE

Thanks to the shifts in the human environment in the past 10,000 years, there are now nearly 8 billion people on Earth rather than just a few million as there were at the time of the agricultural revolution. Average life expectancy then was perhaps 20 to 25 years compared to 72 years globally today, and in many industrialized nations average people now live to their early 80s.[1] Despite this wonderful advance, most people in those rich nations are not as healthy as they *could* be. We must always remember that life expectancy is not the only measure of wellness and that life expectancy, although high, could be even higher. The evidence we have now, as we have seen, is that *in part* the health deficits trace to peoples' failure to chew enough and to maintain a resting oral posture in which the mouth is closed and the upper and lower teeth are lightly touching.[2] Resting the mouth correctly, we emphasize, could in itself reduce many health problems that increasingly afflict industrialized society, even among adults, as well as in the formative years. Other factors of course are involved and can influence peoples' lives directly and, through their influence on appearance, also indirectly.

Peoples' faces often tend to reflect underlying jaw-related illness. Of course, appearance isn't just important as a sort of visual thermometer that can measure the state of our health. Even if most aspects

Image 21. This is a girl with good oral posture. She will likely grow up to be a healthy beauty.

of our looks—size, posture, fitness, beauty, and so forth—do not reflect our key qualities as human beings—bravery, intelligence, compassion, empathy, sense of humor—our appearance, our attractiveness, has enormous social consequences.

Biological visual cues as social indicators are common in primates, the genital swellings of female chimpanzees in heat being a classic example. But there is little sign that one male chimp could improve his dominance over another by being "better looking." He's dominant because he is big, "kicks ass," and is good at building coalitions with other males. Likewise, we doubt if our human ancestors a million years ago had our sense of beauty on which to base mate selection. We can probably thank instead the Great Leap Forward for the origins of appearance's importance in our culture—not as an indicator of health but as part of the cultural evolution of human esthetic tastes. For the first time people could discuss, and gossip about, each other's looks. Tacit social agreements about attractiveness could be reached, and those agreements could change in response to the looks of prominent individuals—just like today. In other words, human taste of what is beautiful and attractive changes over time and differs across cultures.

Although we lack a written record, to say nothing of paleo-TV archives, it seems likely to us that around 20,000 years ago or more appearance started to become a significant factor in cultural evolution. One strand of the evidence is that modern people find Lascaux cave paintings and earlier works[3] created by our ancestors that far in the past to be esthetically pleasing. Of course, the motivation of the artists is open to varied interpretation. Quite possibly the art was originally related to shamanism—attempts to interact in a trance state with an imagined spirit world,[4] just as much of medieval art and the impressive vestments of religious officials were intended to promote in their viewers a sense of communion with a Christian's imagined spirit world. Today people often use their clothes, especially uniforms, including the white uniforms of doctors and dentists, to signal a connection with something bigger than themselves. Another strand of the evidence for the long-ago rise in the importance of appearance is the manufacture, around 75,000 to 100,000 years ago, of the first body decorations in the form of shell beads.[5] We thus bring Stone Age esthetic senses to a world of omnipresent digital imagery that greatly amplifies the ancient importance of how one looks.

In a highly visually oriented culture such as ours, it's only natural that from an early age many of us are concerned, sometimes even obsessed, about our appearance and that of our children. Current cultural standards of beauty and handsomeness are put before us—or thrust on us—almost constantly. Of course there is much subjectivity in ideas of appearance—beauty, cuteness, and so on. Nonetheless, when it comes to faces—the focus of our concern in this book—certain features are generally considered attractive in our culture (for example, balanced features, nose not "too big" in relation to lips, and so on) or unattractive (such as a receding chin). And our cultural standards seem to be becoming ever more dominant globally. For example, studies are now starting to consider the standards of white Westerners (Americans especially) "mainstream," and there is "a growing trend among Asian women to be desirous of a body type promoted by western standards of beauty."[6] This urge is so strong as to cause some Asian women to suffer eating disorders and "poor body image."[7]

We will be using the words *attractive* and *unattractive* as a shorthand for talking about current dominant cultural conceptions of appearance—of facial structure, jaw position, and the like. They are certainly not intended as value judgments about the attractiveness of the whole person. That, we reiterate, depends not only on total physical appearance, but also personality, ethics, intelligence, voice, athletic ability, and so on. It's often thought that basic facial structure is simply a production of genes and little open to environmental influence growing up. But certain aspects of facial and jaw structure are surprisingly plastic, especially during development, and are shaped by cultural habits—how we eat, whether we breathe mainly through our nose or our mouth, how we position our jaws and tongue when not eating or talking, and so forth. In concert these can significantly affect our facial appearance.

One of the jobs of our lower jaw and tongue is to promote the gradual upward and forward growth of the upper jaw so that it does not drop down and back. If kids habitually let their mouths hang open and chew mainly soft foods, their jaws cannot develop properly and will end up too far back, leading to a receding chin and, as we've seen, potentially encroaching on the size of their airway.

It may surprise you that a child not chewing hard enough[8] and habitually mouth breathing or hanging her mouth open at rest can have dramatic effects on physical appearance and general health throughout life. But the evidence that these habits can have such influence is substantial.[9] Mouth breathing by your cute 5-year-old can lead to a 50-year old with crooked teeth, too little oxygen while sleeping, and perpetual exhaustion.

Mouth breathing can change the shape of a child's face and alter its appearance because the jaws are still growing. Results can include long, narrow faces and mouths, less defined cheekbones, relatively small lower jaws, and "weak" chins. Other facial symptoms may include smiles that reveal a lot of gums and, of course, crooked teeth.

Poor overall posture is becoming the norm in industrialized society.[10] Global figures are hard to come by, but a detailed study of 3,520 Czech school children between the ages of 7 and 15 diagnosed 38 percent as having poor posture.[11] The tendency, after much attention by doctors, teachers, parents, and fashion gurus to "good posture" a century or so ago,[12] is now for most people to slouch, and many or most people to do some mouth breathing when they are at rest. Many don't keep their body muscles in good tone. Partial contraction should be the normal resting state of skeletal muscles; that makes them ready to act and helps maintain posture.[13] Those

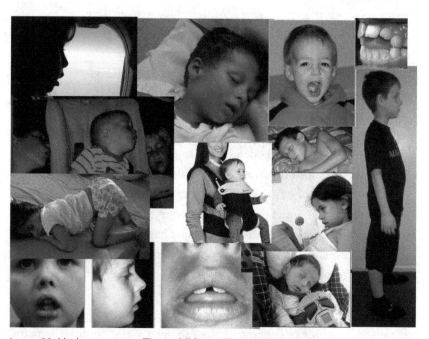

Image 22. *Modern postures*. These children will not grow straight, and neither will their teeth.

who don't maintain tone well will likely struggle with chronic disease, such as back and knee pain as they get older.

Lazy modern body posture exacerbates the problem of poor muscle tone; similarly, lack of jaw muscle tone at rest during a child's growing years tends to stunt growth of the jaws and results in what many would consider a less-attractive face. The exact degree of connection between overall posture and how one holds one's head[14] and oral-facial health remains uncertain, but it seems very likely that slouching can have negative effects on jaw development.[15]

"Cuteness," one of those useful but tough-to-define words, is a term we tend to reserve for young children with certain characteristics adults find attractive: proportionately small nose, large eyes, high forehead. They are similar to the appearance human beings often perceive to be attractive in the young of other mammals—"cute" puppies or kittens, or Mickey Mouse, for example—but there is great variation in patterns of changing attractiveness in our species, and our reactions to it.[16]

Most nonhuman animals lose that human appeal as they mature, as do many human children. As Image 23 shows, poor oral posture and mouth breathing may make previously "cute" children look quite noticeably less attractive as they grow older. These habits often lead to excessive lengthening of the face,[17] a change of appearance dramatic enough[18] to have been christened "long face syndrome" by dentists.[19]

At one time the syndrome could be advantageous. Note the portrait of the Duke of Wellington, the famous British general who was the victor at the battle of Waterloo in 1815, ending the career of Napoleon. Wellington was a great hero to the British and an example of an upper-class man,

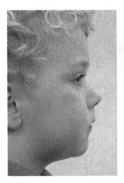

Image 23. Poor oral posture may make previously "cute" children look less attractive as they grow older. (Courtesy of Dr. John Mew.)

Image 24. Poor oral posture and mouth breathing may lead to a change of appearance as the child grows. This is known as long face syndrome by dentists. (Courtesy of Dr. John Mew.)

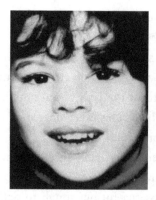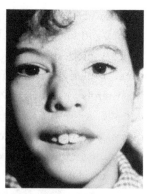

as you can see in his portrait, with long face syndrome. He was admired for his "Roman" nose. His jaws and teeth were set back in the face, which is why his nose and chin looked prominent. A survey of the portraits of the era shows the high frequency of the long face syndrome among the wealthy. This can be seen clearly in the two portraits of Margaret Theresa of Spain, one as a child and one as a long-faced adult.

Centuries ago a prominent nose could be an asset because of its association in parts of Europe with character and aristocratic breeding. Wellington illustrates that all was not lost with poor jaw-facial development in those days—he won many battles, became prime minister, did very well with the ladies, and died in bed at 83.

Image 25. A number of prominent historical figures developed the long face syndrome. For example, the Duke of Wellington (a), shown in an 1804 portrait by Robert Home, shows the set-back teeth and jaw that make his nose and chin prominent. The Infanta Margarita Teresa of Spain shows normal facial development as a 5-year-old child in "Las Meninas," painted in 1656 by Diego Velazquez (b, left). By the time Margarita was a young girl she had developed long face syndrome, as shown in a portrait painted between 1662 and 1664 by an anonymous artist (b, right).

As the example of the Duke of Wellington suggests, appearance in any era is somewhat subjective, despite periodic attempts[20] to establish enduring standards of beauty. One person's "handsomeness" can be another's "plainness," and cultural standards of beauty can be notoriously plastic. A classic example can be found in differing cultural views of "obesity." It carries connotations of unattractiveness and lack of self-control now in Western societies, but a variety of cultures consider it a sign of beauty, and in Pacific societies special efforts were traditionally made to enhance body size. When Westerners first visited Tahiti, they reported that young Tahitian men and women were selected and taken to a special place where they were fed gigantic amounts of food to make them fat and thus, in that society, more sexually attractive.[21] Even as their societies acquire dominant Western attitudes by acculturation, obese individuals in the Pacific still do not have a negative view of their size.[22]

Interestingly, the Tahitian ritual was also designed to lighten skin, its color being a prominent feature of the face. It is ironic that in Polynesia, Japan, and in the unhappy past among some African Americans,[23] a lighter facial skin has been seen as more attractive, whereas light-skinned people in the West often struggle to get year-round tans at some risk to their longevity. Despite such variation, certain facial characteristics may still be thought to enhance or reduce attractiveness in virtually every society.[24] Evaluation of that attractiveness (especially as related to symmetry) to see whether there are universal elements is tricky, and its evolutionary significance is still debated.[25] Scientists have argued, for example, over whether there is natural selection for certain appearances that make some individuals more likely to attract lovers and thus reproduce more.[26]

There has doubtless been some evolution in our concept of what's attractive, especially in the lack of perceived attractiveness of individuals far from the human average, and more attention paid to the appearance of females than males virtually everywhere.[27] There are also fairly consistent uniformities in studies showing that average features, youthfulness, symmetry, and perceived extent of masculinity or femininity of the face are all involved in assays of attractiveness.[28] And there are special brain areas involved in evaluating attractiveness that are different from those brain areas assigned the function of recognition of individual's faces.[29] Thus, developing in an unusual environment that alters jaw configuration can, through changed appearance, affect many aspects of our lives.

But evolved signals may function at much less obvious levels. For instance, subtle changes in attractive female facial characteristics may actually signal the fertile part of their reproductive cycle.[30] Studies also show that people often choose spouses partly on the basis of physical characteristics, including their view of facial attractiveness.[31] That's why we're all blessed with attractive partners!

Despite the debates and some pretty silly ideas about the control natural selection has over various aspects of "attractiveness,"[32] judgments of facial beauty of course continue to play an important role in the lives of those of us who live in industrial societies. Although attractiveness is time and culture dependent, the situation today seems novel. In our heavily visual society with highly photo-illustrated publications, commercial TV and TV commercials, YouTube, videos, selfies, FaceTime, and the like, one is inundated with notions of facial as well as overall "good looks." Many of the people presented as exemplars of attractiveness, whether male or female—movie stars, models, newscast "anchors"—have strong jaw lines and are relatively slender. Maybe the increasing universality of notions of attractiveness may in the end be a tool to help end the oral-facial epidemic.[33]

There is substantial evidence that people judged by Western and Westernized societies (and by *their standards*) to be "handsome" or "beautiful" do better socially (and maintain better health) than those considered less attractive.[34] They tend to be treated differently as children and are less likely to be bullied.[35] They get more votes in elections,[36] make more money,[37] and may be healthier or gain other advantages.[38] They are likely to be treated more leniently by juries.[39] There is also evidence that judgments of attractiveness are fairly consistent among different evaluators[40] as well as cross-culturally.[41] And even babies as young as 6 months show signs of recognizing the cultural standards related to the attractiveness of adults.[42] They looked longer at photos of people judged to be more attractive by adults. Here, as with adults, "averageness" seemed to be a major factor.[43]

We usually recognize very quickly when people in our society will be judged "attractive"; certain characteristics jump out: high cheekbones, strong jaws, defined bone structure, straight teeth, broad smiles, and unmarred skin. Reproduction is a big part of our mental concern; heterosexuals want to choose a reproductive partner who is healthy and judged to be "handsome" or "pretty." The evolutionary "just-so story" behind this is that there are genes that drive us subconsciously to seek such a mate because he or she has a good chance of producing offspring who are also healthy and attractive and will themselves be successful reproductively. It seems logical, but there is little scientific evidence in support of it. The case for normal jaw structure and sexual attractiveness is stronger.

Facial structure that is generally considered attractive usually means every feature appears to have its place in its space and does not deviate much from the average face to which an individual has been exposed.[44] But the recent trend to pay inordinate attention to the appearance of a subset of the population, actors and models especially, as featured in the ubiquitous visual media, may obscure a near society-wide decline in general attractiveness as currently viewed in the West. This is most obvious in the obesity epidemic, where attractiveness and health are often related and where extreme slimness is often considered an ideal of feminine attractiveness. In the view of those who, like Sandra, are trained to identify problems in oral-facial development, there is a building health epidemic signaled by reduced facial attractiveness, as conventionally defined, just as there is in body weight. This could mean that, if the oral-facial epidemic continues, humanity is in for a substantial shift in what is considered attractive, one severe enough that it will become obvious even to laypersons. When the teeth are crowded, there is a collapsing of the face where its basic profile becomes more concave and the structures that we find attractive don't fit. We see retreat of cheekbones, softness in the jawline, crooked teeth, and breathing through the mouth instead of the nose—all because developing the face in the new industrial environment has resulted in jaws with too little room for the teeth or for the tongue.

When habits detrimental to health are socially acceptable, even actively encouraged, behavior is hard to change. Smoking was so "normal" and heavily advertised that during World War II a soldier's K-rations (a compact box of food issued to those in combat) contained a pack of ten cigarettes. Paul and his friends in the 1930s picked up and smoked butts from the street to look "grown up" (and enjoy the pleasures of nicotine addiction). They already called cigarettes "coffin nails" even at 7 or 8 years old, but social pressures on the young to smoke were high. In contrast, smoking

Image 26. Notice the contrast between Jay Leno and Robert Redford. Leno's profile is dished, probably resulting in crowding of his teeth and tongue. Note the more forward aspect of the center of Redford's face, probably resulting in a more roomy mouth.

is looked down on today in the United States and many other countries. America recognized the public health consequences of the smoking epidemic by the 1980s and gradually managed with laws, restrictions on use and access, and resultant changes in social norms to do quite a lot about it.

Our society has begun to recognize and work on the serious health issues surrounding the obesity epidemic, but here the causes are less broadly recognized, and people still tend to see it, falsely, as traceable to individual failing—a "lack of self-control." Despite the massive health hazard posed by obesity, even legislation simply designed to limit the size of sugary soft-drink bottles has proven difficult to pass.[45] The pleasures to those hooked on certain industrialized diets and the financial benefits to those who purvey those diets so far have tended to prevent much meaningful action on the well-recognized epidemic of obesity—small wonder the oral-facial health epidemic remains hidden.

As we have seen, facial deformity, as measured by malocclusion, is a rampant condition, but it is no more widely recognized now than it was in Wellington's day. The costs of a cure are relatively high, and the obvious immediate benefits relatively few. The sometimes intimate relationship of crooked teeth to serious illness is basically unknown to the general public. Furthermore, the food industry, and in particular the baby-food industry, is not perceived to be an evil monster like the tobacco industry.[46] If the social knowledge of the importance of chewing to development were readily available, baby-food purveyors could easily provide more chew-promoting products. Most people easily realize they were not born to be deformed but to have balanced features, ample airways, and attractive appearances. As we have said, the epidemic can ultimately be traced to the hunter-gatherer genes now finding expression in industrialized societies, an environment quite different from that of our ancestors. When we do recognize persons suffering from poor oral-facial development, we often choose to accept it as an accident without thought to how specifically it occurred or how they might be helped.

It is our hope that recognizing facial deformity as a preventable condition will lead to a social-political movement to change that environment to prevent the condition. We especially hope that health professionals and caregivers become much more broadly aware of those risks and that medical and dental professionals revise their training in facial-oral health issues to make prevention and treatment readily available. Prevention is the key, and it needs to involve not just health care workers but attitudes in society as a whole. If the ability to recognize the need for treatment were to become widespread, treatment when needed could start for many more 5 to 7-year-olds, when it would be most effective.

In one way it could be a blessing that attractiveness can be a thermometer that can signal underlying serious disease. Extreme malocclusion should be a symptom of great concern, but the more subtle distortions caused by poor jaw development, and obvious mouth breathing, if widely understood by the public as symptoms of likely underlying problems, should help get many more youngsters treatment in time. Indeed, by their very commonness, when pointed out, they may encourage broad changes in social practice.

CHAPTER 5

DEVELOPMENT
AND ORAL POSTURE

So where are we? Many if not most people in the environments of in-dustrialized societies no longer develop their jaws to their full potential size, which would allow them to have their full complement of 32 teeth in alignment. They typically don't provide an eating environment for their children in which the jaws fully develop[1] or sufficiently encourage oral posture that is best suited for their children's development. As a result, the jaws may not develop enough room for all the teeth, and the tongue may start sitting lower in the mouth. The lower jaw will then tend to hang open at rest, and the airway may become constricted. The optimal oral posture, we reiterate, is when the lips and jaws are closed when the mouth is resting, with upper and lower teeth in light contact and tongue pressing against the roof of the mouth.[2] Poor oral posture can be a consequence of breathing difficulties, however. If there is a nasal blockage from an allergy or airway obstruction from enlarged tonsils and/or adenoids, more frequent mouth breathing is the typical result, exacerbating the underdevelopment of both the oral cavity as well as the airway[3] and increasing the likelihood of maloc-clusion.[4] As just one indication of these effects, restoration of nose breath-ing after removal of adenoids in 7- to 12-year-old Swedish children led to more horizontal (natural) growth of the jaw.[5]

What are the origins of the oral posture part of the development prob-lem? The problem's very existence tells us that no intelligent being designed our jaws to be too small for our teeth and tongues. The evolutionary story is, by contrast, much more logical than the "intelligent design" view and tells us a lot about ourselves and how our past triumphs have helped set the scene for the array of present issues we're discussing. The appearance of spoken complex language was a critical step in making *Homo sapiens* the

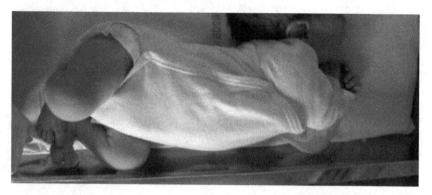

Image 27. A serious problem with children with inadequate airways is that it can lead to SIDS (sudden infant death syndrome). The parents of this child saw their baby in a funny pose and took this picture, not realizing that the infant was trying to open its airway. After the child's tragic death, they showed it to their physician. (Courtesy of Christian Guilleminault.)

dominant animal[6] on Earth today. It was a great evolutionary triumph, but it required crucial changes in our ancestors' facial–airway configuration that, as an unfortunate byproduct, have made people more subject to a variety of serious problems, including those in oral-facial development. The critical change is thought to have occurred in the last 150,000 years, perhaps as recently as some 50,000 years ago,[7] sparking "the Great Leap Forward"[8] mentioned in Chapter 3. The development of speech with syntax (the ability to put words together in a meaningful order to form sentences) led to the ability of our ancestors to plan and execute complex strategies and discuss imaginary situations.

Although chimpanzees, our close evolutionary relatives, can have quite complex hunting strategies, no chimp could communicate something like: "Herman, you go around the left side of the hill, and I'll go the right. If we've got the bear between us, attack only if you're sure you have a clear escape route." No gorilla can discuss "counterfactuals," such as "What if I hadn't tried to beat up that other silverback male?" So the anatomical difficulties accompanying the evolution of this capacity were clearly less important than the new communication ability.

The anatomical evolution—the change in the DNA that made complex speech possible (along with the huge evolutionary advantage this brought in its wake)—at the same time narrowed the airway (a relatively small disadvantage).[9] So at least some part of our dental-airway problems goes back to when our distant ancestors developed the capacity to talk[10] and communicate in more than a series of grunts and gestures. It involved complex changes in human throats. The larynx (familiarly known as the voice box) at the top of the tube leading to our lungs dropped, making a larger air space (technically known as the supralaryngeal vocal tract, or SVT) above the larynx and behind the tongue. That space can be used to greatly modify the sounds our exhaling air can make.

The ability to clearly enunciate vowels and consonants has huge advantages, but like many evolutionary advantages it carried disadvantages as well. As the advantages of walking on our hind legs are paid for in back pain and hernias, those of being able to whisper sweet nothings to a prospective lover are paid for with a higher probability of choking to death. That's because, with the dropping of the larynx, air and food travel down the same pipe in the neck, with a system of flaps and valves that shuttle air toward the lungs and food toward the stomach. Sometimes the system malfunctions; food is shunted toward the lungs, obstructing the airway. (This malfunction led to making American thoracic surgeon Henry Judah Heimlich famous for the emergency unblocking procedure named after him.) Infants are spared this threat because the dropping of the larynx and forming a single food–air channel doesn't occur until a child is about 2 years old. That keeps the pipes separate and allows a nursing infant to suck and breathe simultaneously without choking. But unless the complex air

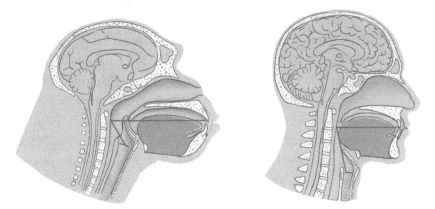

Image 28. In the speechless chimp on the left, the tongue rests completely with in the jaw with its back far from where air must pass to the lungs. As a consequence of acquiring speech, the back of the tongue easily can impinge on the airway, making humans, on the right, more vulnerable to OSA.

intake anatomy changes to allow speech, involving such things as evolutionary shortening of the jaws and changes in their relationship to the base of the skull, develop properly, adults can end up in deep trouble with their nighttime breathing. The epidemic of obstructive sleep apnea thus may trace in part to our superb ability to communicate.[11] Of course the sleep apnea epidemic is of much later vintage than the development of speech, so the anatomical shift that enabled human vocal language is by no means the whole story.

How much and what kind of work our muscles do are also critical environmental factors in oral development. As Dr. Suzely Moimaz and her group of specialists in childhood dentistry put it: "Breastfeeding is seen as a determining factor for proper craniofacial development, because it promotes intense exercise of the orofacial muscles, favorably stimulating the functions of breathing, swallowing, chewing, and speech production."[12] Other scientific literature shows lack of breastfeeding correlates with underdevelopment of the jaws, increases in mouth breathing, and consequently with more malocclusion.[13] Breast milk, you'll recall, also may be helpful in strengthening immune responses and reducing the runny noses that also contribute to mouth breathing and, through its effect on jaw development, crooked teeth. Not only is breastfeeding helpful in avoiding malocclusion, but pacifier use, in contrast, encourages the development of malocclusion,[14] likely because it reduces the amount of breastfeeding.[15] Bottle feeding had a similar but less pronounced negative effect, as for example in a study of the feeding patterns of over 1,000 Italian preschool children.[16]

The whole issue of bottle feeding began to appear rapidly after the mid-18th century in Western Europe, especially in England and France. It was then, as pediatric dentist Kevin Boyd of Lurie Children's Hospital in Chicago has pointed out, that women began to enter the textile mill workforce in large numbers. In the initial decades most of the females working there were children or single, but industrialization led to the abandonment of thousands of centuries of patterns of prolonged infant/childhood nursing and weaning. Rather than working at home in a cottage industry and being able to nurse their babies "on demand" for years, the era of bottle feeding dawned for working women with artificial nipples, pumped breast milk, infant formulas, and pacifiers.[17] The result was the loss of the nursing environment so essential to the development of normally large, forward-thrust jaws. In contrast to bottle feeding, babies being nursed naturally perform intense muscular work to suck the breast, tiring the muscles working the jaws, which causes the child to sleep and does not require use of pacifiers, fingers, and other objects. The bottom line is that sucking activities that do not provide milk—pacifiers, fingers, and so on, create malocclusion.[18]

Children today are seldom taught to chew their food thoroughly (20 times per mouthful), to keep their mouths closed when not eating or talking, and to breathe through their noses even when eating or talking. Strong pressures from muscles doing tough chewing are an important part of the developmental environment, influencing how the jaws and face are formed.[19] Equally or possibly more important are the weaker but more persistent pressures from muscles that keep lips and jaws closed at rest and the tongue against the palate. These gentle pressures cause the tongue, teeth, and palate to interact with each other and shape the living bone into the patterns set by millions of years of genetic evolution. *In sum, poor oral posture, like less-determined chewing, disrupts the environment-gene interactions that would lead to optimal size and configuration of the jaws and airway.*

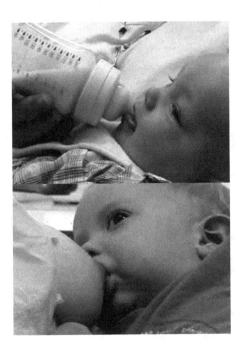

Image 29. Babies who are bottle fed receive milk passively (top) whereas babies who are breast fed have to perform muscular work to successfully feed (bottom). (Courtesy of Dr. John Flutter.)

Although such lasting influence from slight muscular pressure may seem counterintuitive, that gentle muscle action exerted consistently over a long period can have a major effect on the teeth and bone. That has been demonstrated by both observations on people and experiments on monkeys.

Open-mouth habits as a source of the recent dramatic reduction in human jaw size have been suggested by experimental work on rhesus monkeys. Scientific ethics have properly limited the amount of experimentation that can be done on human subjects, and the use of other primates in experiments is happily under increasing ethical scrutiny.[20] But there's much to be learned from previous studies of the influence of mouth breathing on jaw size in rhesus monkeys. There is every reason to believe their developmental systems are extremely similar to ours, so these studies can tell us a lot about what happens in human beings, where there is considerable evidence of the influence of mouth breathing on jaw structure[21] and sleep-related health problems.[22]

In the 1970s pioneering anatomist Egil Harvold carried out a series of experiments on rhesus monkeys. The experiments showed,[23] for example, that when rhesus monkeys between 2 and 6 years old were forced to become mouth breathers by having their nostrils blocked with silicon plugs, they first showed a wide variety of behaviors with their tongue, lip, and jaw position, behaving "erratically" as the animals adjusted to mouth

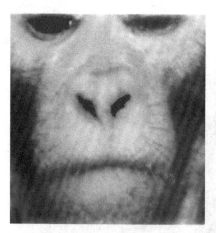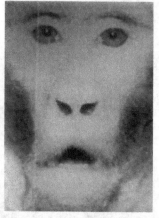

Image 30. In Egil Harvold's experiments, the monkey on the left was a "control" and the one on the right had its nostrils blocked. The hanging open of the mouth caused by the blockage produced narrowing of the jaws and the long face syndrome, as seen on the right.

breathing. The mouth breathers developed longer faces with more downward sloping mandibles. They also developed forms of malocclusion that varied substantially from individual to individual. Apparently, as is clear in human beings, differed solutions for coping with the novelty of mouth breathing generate different pressures on the flexible jaw development system.[24] All, however, showed a narrowing of the dental arch in the lower jaw, and a shortening of the arch of teeth in the upper jaw. The result was an "incisor cross bite," a form of malocclusion with the incisors of the lower jaw in front of those in the upper jaw. When the plugs were removed after 18 months, most of the changes tended to reverse. Harvold also showed in the rhesus that gentle pressure of the tongue against the palate causes the facial skeleton to grow in a way that produces a wide palate and a normal arch of teeth.[25]

There are, of course, differences in the mouth and facial anatomy of rhesus monkeys and human beings, but Harvold's experiments reinforce the strong evidence that breathing patterns can change human facial development and that mouth breathing can lead to the "long face syndrome,"[26] which many now consider culturally unattractive.[27] Long face syndrome is correlated with a pattern of keeping the teeth in upper and lower jaws about 2 millimeters apart.[28]

Harvold's work illuminates what we said was probably an important basic contributor to difficulties with oral-facial health: humanity's shift to indoor living. That moved people into a situation reminiscent of the blocked nostrils suffered by Harvold's monkeys: an allergen-rich environment ideal for producing stuffy noses. This shift and the accompanying exposure to allergens, like learning to grow and process softer foods, is another aspect of humanity's distant history that makes kids today more likely to have allergies and to develop jaw problems.[29]

When babies are born they usually breathe through their noses except when they are crying; some cannot breathe through their mouths even when it is appropriate—newborns with rare congenital blockage of the nose may suffocate rather than breathe through their mouths.[30] Blockage of the nose at an early age is frequently due not just to regular colds but, according to the "hygiene hypothesis," also due to living in enclosed, allergen-rich spaces,[31] while not having immune systems well "trained" by close contact with farm animals, infections, and "dirt" early in life.[32] It seems ironic, but the modern "clean" environment[33] seems to have deleterious effects on the immune system's ability[34] to prevent allergic attacks and asthma on the one hand[35] and on the other to concentrate allergens indoors (such as dust mites and their feces) and generate novel ones (like formaldehyde).[36] That modern environment might possibly include comparatively little breastfeeding,[37] which can pass immune protective effect from mother to infant. Stuffy noses in kids may be one reason why mouth breathing and malocclusion are now so widespread and common.[38] Today's conditions in indoor environments may not be nearly as harmless as people often assume and instead are causing chronic disease.[39]

To see how important this could be, consult the pictures of the unfortunate boy who developed an allergy to his pet gerbil (Image 31). The advantages of good tight shelter presumably has had an unfortunate side-effect of altering the environment in a way that sometimes hinders oral-facial development.

The importance of keeping the oral-facial environment appropriate, especially for young children, can't be overemphasized. Interestingly, human patients who have undergone jaw surgery to correct a skeletal deformity will relapse if they are not trained to maintain a suitable environment—proper oral posture and nose breathing.[40] This is testimonial to the plasticity of living bone, something on which the techniques of orthodontics, which move teeth through bone, depend.

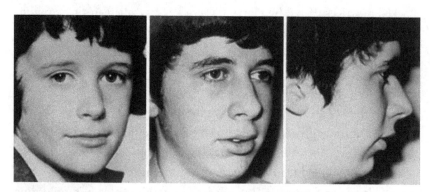

Image 31. Allergy can block a young person's nostrils as thoroughly as plastic plugs can block those of a rhesus monkey. Look at the consequences here for an attractive youth (left) who got a gerbil for a pet. He was allergic to the gerbil, and the resulting nasal congestion and mouth breathing redirected the growth of his jaw with sad results (right). (Courtesy of Dr. John Mew.)

Scientists are still learning how important particular environmental influences are to the development of vertebrate—and thus human—jaws and faces.[41] We suspect that scientists will increasingly discover that cultural changes in children's earliest environmental influences (including returns to cultural patterns associated with long-gone environments!) will give the best results in combating the epidemic of oral-facial problems. This could mirror what is known about the development of some other human systems—that there is an early "critical period" during which the environment *must* be appropriate for normal development to occur. A classic case was the demonstration that unless very young children are exposed to a normal visual environment, they may never learn to see as well as people sighted since birth. Children born blind who had their sight restored surgically after this critical period had passed did not develop normal vision. For example, one such person could perceive a dark object against a light background but couldn't discriminate between a cross and a circle and reverted to the life of a blind person, not using vision to orient to the world.[42] In another study, a person blind from the age of 3 until 46 still had not developed normal vision two years after surgical restoration of sight.[43] There is variation here, though: another person with sight restored did develop enough visual acuity at least to function adequately in society.[44]

Similarly, for ideal jaw-face-airway development there appears to be a critical period in the first decade (and perhaps more so in the first year or so) of life during which behavior must include proper oral posture (keeping the mouth closed), consuming "chewy" foods during and after weaning, and other factors we've discussed.

After having poor oral posture during that critical period, returning to a normal path of jaw development has proven much more difficult, although there have not been scientific studies of the degree of success that can be achieved and the individual variance in it. The situation seems roughly analogous

Image 32. Boy being weaned to hard, chewy food. His first bite, at 6 months, at a raw pear.

to that of a critical period of language acquisition in the first few years of life,[45] though a few people retain the ability to pick up new languages effortlessly far into maturity.[46]

Writing of the critical period in oral-facial health, John Mew stated:

If during infancy the maxilla (upper jaw) receives insufficient occlusal (teeth in touch) and/or lingual (tongue) support it tends to drop down. The age of eight seems to be crucial as beyond this age the maxilla becomes progressively more firmly attached to the basi-cranium (base of the skull). By puberty it is relatively immobile unless moved by active appliances but the mandible remains adaptable for some years after this, particularly in boys.[47]

In any case, clinical experience and first principles (in any unfolding complex system early intervention is more likely to effect major change than later intervention) make it clear that jaw development will be more easily influenced in youngsters than in oldsters.

For millennia hunter-gatherer women breastfed their children for years, eventually weaning them on the tough, chewy foods they themselves ate ("baby food" was yet to be invented). Human developmental systems evolved so that the results of this regime were a typically healthy pattern of skull and jaw formation. Jaws, as we saw in Chapter 1, were relatively large and roomy, with teeth set out in neat rows, not overlapping or pointing in weird directions. It is unlikely to be coincidental that today breastfeeding is associated with reduced malocclusion.[48]

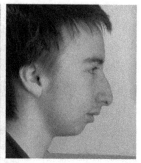

Age 5   Age 17

Image 33. Parents always have extremely cute young children but should learn to recognize the early signs of misdirected growth. See how the boy on the left is good looking but hangs his mouth open; the results are on the right. (Courtesy of Dr. John Mew.)

With the evolution of agriculture, cultural patterns of food production and consumption began to change,[49] with the first changes likely more in the techniques used (gathering versus growing grains) than in the types of food consumed.[50] The patterns of change were complex, determined in part by such things as differences in the actual food consumed, shifts in some areas from hunter-gatherers heavily dependent on marine resources to primary dependence on cultivated plants, the kinds of preparation (dicing, grinding, fermenting, roasting, boiling) used, and so on.[51] It seems likely, however, that the great reduction in chewing and increase in stuffy noses associated with the modern malocclusion epidemic came on much later and gradually. Most of the literature on dental transition that accompanied the development of agriculture focuses on the increase in cavities (caries) associated with moving to higher-carbohydrate diets and mouths much more favorable to decay-causing bacteria, and differences in wear as a result of stone fragments in grain ground in mortars. There is relatively little reference to the incidence of malocclusion.[52]

Softer foods that children could easily eat without much chewing became available with a more sedentary lifestyle and, likely, more time for cooking. After hundreds of thousands of years, the wide adoption of cooking had reduced the need for the long jaws and intense chewing that were required to extract nutrients from raw food.[53] One result of the agricultural revolution, added on top of the earlier, trickier airway design that developed in support of the evolution of language, was earlier weaning made possible by the availability of softer cooked foods. That led to distorted muscle use as the complex muscular sucking of nursing was replaced by much simpler motions. This changed the basic patterns of oral development. So might have the gradual adoption of diverse eating utensils. Use of spoons, forks, and chopsticks could become common cultural practice as people stopped continually moving and no longer needed to carry all their possessions with them.[54]

Eventually, in the last few centuries, all those new influences on oral development contributed to a rapidly rising incidence of too-small jaws and badly fitting teeth. Collectively they appear to explain the increased need to remove wisdom teeth, mouth breathing, problems with swallowing, and sleep apnea, as well as some difficulties with speaking (which depends on proper control of the tongue and other muscles of the mouth).[55]

These effects are related to the soft foods onto which young infants were increasingly weaned at a critical developmental period,[56] especially after the industrial revolution with the invention of a "baby food" business. *Indeed there is evidence that human jaws got smaller between medieval and modern times as the transition toward the industrial diet accelerated.*[57]

Interestingly, as with the Duke of Wellington (Chapter 4, Image 25a), the likely effects of too soft a weaning diet and subsequent consumption can be seen by comparing the facial structures (and related airway openness) of rich and poor Westerners from a few centuries ago. The rich and their babies had much more "delicate" diets and presumably chewed less as small children. The jaws of the rich, one can assume from their portraits, tended to be underdeveloped, palates narrowed, and the airways more compressed. This can often be detected in painted portraits of the rich whose noses tended to be hooked down, the foreheads sloped backward and the space between the tips of the nose and chin seemed elongated. Apart from their other problems, poor people in those days were much less likely to suffer distorted faces and jaws from too soft diets.

In the West at least, people today are almost all moving to what once was a rich person's diet. The dropping of the larynx for speech, the shrinking of the oral cavity, and placement of the back of the tongue adjacent to the airway, combined with the modern lack of tough chewing by children, modify the development of the entire lower part of the skull and the jaw in ways that allow the back of the tongue to spill into the airway and partially restrict it. Mouth breathing is a compensation mechanism. When the mouth opens, the tongue is able to move forward away from the breathing tube, and breathing is easier, but the advantages of passing the air through the nose, described in Chapter 6, are lost. And, as we have seen before, underdevelopment of the jaws and vertical growth of the face results.

To repeat, modern environmental trends, from slurping down fast foods to living indoors with dust mites, have led to problems in the development of our jaws and faces. As we've seen, growth patterns depend both on how we use our jaws and also, surprisingly, on how we *rest* them. Grandmothers knew intuitively that rest was critical to our growth: "you need to sleep in order to grow"; indeed, science has now shown that growth hormone is secreted most efficiently between 11 pm and 2 am.[58] The release of that chemical messenger is dependent on our circadian rhythms, the roughly 24-hour cycles in our body's processes in response to the alternation of day and night.[59] Folk knowledge tells us that we need to sleep to be refreshed and at our peak potential in mind, body, and spirit. Sleep for children used to be given priority over many activities, and people knew that the number of hours that they slept was important, but they did not (and most still do not) focus on the *quality* of their sleep.

CHAPTER 6

DISORDERS OF
BREATHING AND SLEEP

Disruption of your breathing while you sleep can have nasty effects on your health,[1] including heart health, diabetes,[2] stroke, and mental disorders.[3] Children with obstructive sleep apnea (OSA) are more likely than other children to have poor learning skills, behavioral problems, ADHD, brain injury, and depression.[4] One of the main symptoms indicating that a child is headed for such problems is mouth breathing.

Mouth Breathing and Hanging Your Jaw

Poor oral posture typically leads, among other things, to mouth breathing—a condition that demands prompt treatment in youngsters.[5] A substantial proportion of people in industrialized societies are mouth breathers.[6] For example, in a careful study of 150 school children between the ages of 8 and 10 in Recife, Brazil, 80 (53 percent) were mouth breathers.[7] The children were carefully observed without their knowledge and were tested by observing the steam patterns when they breathed on mirrors. Investigators also tested the children by seeing if they could hold water in their mouths for three minutes (mouth breathers cannot). These tests would exclude individuals with open-mouth postures who still breathed through their noses.[8] Some people who keep their mouth open still continue to be nasal breathers. For this reason John Mew prefers to use the more accurate "open mouth posture" to describe the problematic condition, but we'll stick with the more familiar "mouth breathing."[9]

The nose is a complex structure with many functions.[10] Air taken in through the nose is warmed, moistened, and filtered, and small amounts of bactericidal nitric oxide,[11] which may play a structural role in maintaining lung health,[12] are added to the air before it goes to the lungs. Breathing through the mouth has none of these advantages and can have unhappy consequences that extend far beyond interfering with development of appropriate jaw size[13] and thus producing crooked teeth—remember Harvold's monkeys! Here's an edited litany of problems associated with mouth breathing from a group of pediatric dentists at the University of Pernambuco:

The most common complaints of oral breathers are: breathlessness or respiratory failure, gets tired easily during physical activities, back or neck pain, olfaction and/or taste impairments, halitosis, dry mouth, wake up choking during the night, bad sleep, day time sleepiness, dark spots underneath the eyes, sneezing, abundant saliva when speaking, among others. As physical consequences, the oral breathing child has many physical traits: long face, dropped eyes, dark spots underneath the eyes, open lips, (sagging) and dry lips, narrow nostrils, (weak) cheek muscles, high palate, narrowing of the upper arch, and (malocclusion). Oral breathing also alters posture, morphology and the tonicity of phonoarticulatory organs (strength of the organs of speech).[14]

Because mouth breathing can negatively affect your oral health, it's important to note some of the signs and symptoms of this potentially dangerous habit. Many mouth breathers in fact don't even realize they're doing it—Paul didn't. Easy to observe symptoms that *may* signal someone is a mouth breather include:

- Dry lips
- Dry mouth
- Snoring and open mouth while sleeping
- Numerous airway illnesses, including sinus and ear infections and colds
- Chronic bad breath
- Swollen and red gums that bleed easily

Mouth breathing can quickly dry out the mouth and decrease saliva production. Saliva is extremely important in neutralizing acid in the mouth and helping to wash away bacteria; without it, the chance of tooth decay increases.[15] Dry mouth is one of the causes of gum disease, cavities, the inflamed gums of gingivitis that may advance to periodontitis, where microorganisms cause pockets to form around teeth, which loosens them from the gums. If left untreated, periodontitis can cause bacteria to gain access to the bloodstream, where they may, for instance, colonize heart valves and cause the serious disease of endocarditis.

Tooth decay can also be traced to mouth breathing and dry mouth.[16] We emphasize that oral infections are not trivial matters. Death is always a possibility, as has happened in numerous cases. One such tragic story was that of Deamonte Driver, a 12-year old boy in Maryland. In her recent book *Teeth*, Mary Otto tells the story of that boy. We know that keeping the mouth open dries the mouth. and less saliva means more cavities. Recalling briefly Deamonte's story brings about the importance of taking prevention seriously, both for the human toll as well as the economic costs to society.

Deamonte came home from school one day, complaining of a head-ache—nothing special. His grandmother took him to Southern Maryland Hospital Center, The doctors gave him medicines for headache, sinusitis, and a dental abscess. The next day, a Thursday, Deamonte went back to school. "On Friday he was worse," his mother Alyce said. "He couldn't talk." Alyce took him to Prince George's County Hospital Center, where her child was given a spinal tap and a CT scan and diagnosed as having meningitis. He underwent brain surgery for an infection on the left side of his brain at Children's National Medical Center in Washington, D.C. On Saturday he started having seizures and was operated on again. In that operation the ab-scessed tooth, a molar on the upper-left side of his mouth, was also removed. This tooth was infected to the core, and bacteria from the abscess had trav-eled to Deamonte's brain. The doctors said he was "fighting for his life."

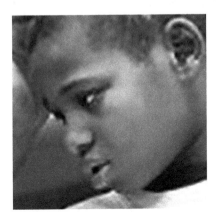

Image 34. Deamonte Driver, who died in 2007, aged 12, after an infection from an abscessed tooth spread to his brain. Sadly one might conclude from the available photographs, that Deamonte was a mouth-breather or had a habit of hanging the mouth open. (Photograph: *Washington Post/Getty*.)

He slept for two days while his family gathered around the bed and prayed. Deamonte finally awakened and spent more than two weeks at Children's National; he was then transferred to the nearby Hospital for Sick Children, for an additional six weeks of physical and occupational therapy, where he did schoolwork and enjoyed visits from relatives and teachers from his school. But his eyes seemed to remain weak, and on February 24 he refused to eat but seemed happy—calling to Alyce as she left him to be sure to pray before she went to sleep. She was called the next morning and told that Deamonte was unresponsive. She rushed back to the hospital, but when she arrived he had passed away. "When I got there," Alyce said, "my baby was gone." Deamonte's family was poor, and Alyce had encountered problems finding dental care. For lack of that care, his abscessed tooth had killed him.

Deamonte's case is a good example of the subtle connections between apparently minor issues and sometimes serious, life-threatening disease. A stuffy nose can lead to poor jaw development and mouth breathing. Mouth breathing can lead to tooth decay, gingivitis and periodontitis, and periodontitis can lead to endocarditis. And endocarditis can lead to heart attack or stroke. In children and adults, breathing through the mouth can also lead to poor sleep, lower oxygen concentration in the blood, and a habitual head-back posture to keep the airway open. In addition, in a developing child routine breathing through the mouth may lead to the face growing longer and narrower than it might otherwise, the nose becoming flatter and the nostrils smaller, with the upper lip becoming thinner and the lower lip poutier. Mouth breathing, which may result in hyperventilation (abnormally deep or fast breathing) and hypoxia (lessened ability to deliver oxygen to tissues that need them), may also cause asthma or make it worse.[17]

Statistics are hard to come by, but in 2016 *Consumer Reports* asserted, "Up to 70 million Americans have a sleep disorder—persistent difficulty sleeping and subsequent trouble functioning during the day."[18] Chiming in, *New Scientist* reported, "Poor sleep is a major risk factor for obesity, diabetes, mood disorders and immune malfunction. Put simply, poor sleep can shorten your life."[19] Sleep disorders take many forms, some of which have little or nothing to do with our breathing, airways, and jaw position: sleep disruption brought on by depression, anxiety, pain, too much caffeine, working late on your tablet, and the like. Yet a significant proportion of the disorders do involve our breathing habits and airway size, which, in turn, are often related to our jaw configurations and oral posture.

Sleep-Disordered Breathing, Snoring, and Sleep Apnea

Breathing is our number one priority in life. It is something your brain pays attention to on a continuous basis. When your breath is taken away, all your energy and effort go into finding your next breath. Some sleep disorders are so disruptive that they may trigger just that response. The most severe type we have already met, obstructive sleep apnea (OSA). Some instances of sleep apnea can be traced to a neurological disorder called central sleep apnea, in which you stop breathing because your brain doesn't immediately tell you to continue, but it is rare compared to obstructive sleep apnea. OSA usually involves blockage in the throat itself, and an estimated 12 million people suffer from the disease in the United States alone. A common cause is problematic oral-facial development resulting in a jaw too small for the tongue to rest comfortably inside its confines; in these cases, the back of the tongue may fall into the throat, interrupting airflow. As John Remmers, the Harvard-trained physician who coined the term *obstructive sleep apnea*,[20] said, "Structural narrowing of the pharynx plays a critical role in most, if not all, cases of OSA. This is due to upper and lower jaws being recessed in the face. *OSA would not exist if the upper and lower jaws were ideally placed in the face.*"

Although other factors, such as obesity, overconsumption of alcohol, and tobacco use, can contribute to OSA, as they do to snoring, we agree with Remmers that the basic cause is structural. Remmers also predicted that OSA would become the most common chronic disease in industrialized countries.[21] Although other ailments like obesity and type II diabetes seem bound to give tough competition, despite many millions afflicted few steps have been taken to understand and prevent OSA. As with many of those other chronic diseases that plague us, it is simply another runaway ice cube being kicked under the fridge.

Mouth breathing itself might be an important source of disease. Anything that dramatically disturbs your breathing will be addressed by your autonomic nervous system (the system that does not require you to think before taking action). Such disturbance of breathing triggers the autonomic system's "sympathetic" division to shift into an automatic "fight-or-flight" response—even if the nervous trigger is not the growl of a hungry tiger coming at you but an approaching final exam or fear that your boss won't promote you this year. The minute your breathing is interrupted, you gasp to restore the oxygen flow to your lungs, and the sympathetic nervous system speeds up your heart, jacks up your blood pressure, has you gasp air through your mouth, and diverts blood to your leg muscles and away from organs and processes (gut, reproduction, growth) not immediately needed for survival. You subconsciously prepare to run, fight, climb, or do whatever is required to stay alive. If the tiger doesn't get you, the other division of the autonomic system, the parasympathetic, will help you calm down and restore normal healthy bodily functions. Mouth breathing is obviously a great benefit if you are being chased; it gets more oxygen to your lungs and bloodstream to power *very* hardworking muscles. Growth is turned off when all your energy is channeled into saving your life—allowing for growth another day. As stress expert Robert Sapolsky[22] put it: an antelope fleeing from a lioness doesn't have the time or energy to grow antlers.

Sapolsky is not sure anyone knows whether mouth breathing by itself, or in conjunction with sleep apnea, triggers a partial "run from the tiger" response.[23] But there is no question that habitual difficulty in getting a restful night's sleep is an important stressor and over time can seriously damage your health, leading to increased susceptibility to infections like influenza,[24] gut problems, heart disease, and other ills.[25] Indeed, mouth breathing may also interfere with the normal growth and development associated with nose breathing. A team of Swedish scientists has pointed out that "improving nasal breathing in snorers increases nocturnal growth hormone release."[26] If you can get your kids to switch from mouth to nose breathing, in other words, their growth will be enhanced.

Breathing disruption triggers the same sort of stress response that prepares your body for combat, when it should be enjoying restorative sleep.[27] You're trying to rest, but at the same time you're struggling to supply your brain with the necessary oxygen. In sleep apnea cases, where the intake of oxygen is intermittently limited or nonexistent, the brain responds by shutting down all programs that are not essential to survival. The conflict is not confined to your brain but is carried to every part of your body in nerve impulses and hormonal messengers. By trying repeatedly to make these critical adjustments, the brain is wreaking havoc on itself, creating destruction in a key part involving memory (hippocampus) that can't ever be healed.[28]

Impaired nighttime breathing, along with the other stresses of modern life, can put us in "sympathetic overdrive," which can have disastrous long-term effects on the human body, including heart and blood vessel disease.[29] This apparently happens because sleep apnea elevates blood pressure—good evidence of which is declining blood pressure in people whose apnea is corrected.[30] And if it is discovered that mouth breathing alone, without cutting off the flow of oxygen, can trigger a partial version of this sympathetic response, it will greatly underline the negative results of the habit.

Sleep breathing pattern is important to monitor from a very young age. Normal breathing is quiet and occurs through the nose with the mouth closed. There should also be no excessive tossing and turning in bed, as regular respiration equals peaceful sleep. Any significant deviations from these patterns are considered health threatening and should be dealt with as soon as possible.

Snoring is a sign of a restricted airway, of possible sleep apnea, and can be associated with heart disease,[31] but snorers do not suffer the total interruptions of breathing that afflict victims of sleep apnea. Like sleep apnea, it is common in the modern world, one guess being that about 30 percent of people over 30 snore, and 40 percent of those over 50.[32] In Great Britain the prevalence of snoring was similarly estimated to be some 25 to 40 percent,[33] and in one town well over half of the middle-aged men snored.[34] Snoring is more common in men than in women, in some surveys almost twice as common.[35] In today's society many see it as normal, but the opposite is true.[36] Very long ago, snoring was likely rare, if not nonexistent, among humans. Indeed, in hunter-gatherers it would be dangerous, calling predators' (and enemies') attention to sleeping and thus relatively helpless individuals. In classic times snoring was well recognized, at least as a disturber of others' sleep, but it is a very complex process whose relationship to serious disease, like sleep apnea, is still not fully understood.[37] The changes in jaw development seem to have caused it to become more widespread among humans of all ages. When snorers sleep, the tongue falls back, partially restricting the passageway between their noses and lungs, causing a rhythmic rumbling sound. Snoring is especially common in older individuals, whose loss of muscle tone makes it more difficult to keep the airway open. Frequent intake of alcohol, tobacco smoke, and other drugs, as well as obesity,[38] can also bring on or worsen snoring.

Childhood snoring has become commonplace even in preschoolers.[39] David Gozal, University of Chicago sleep scientist, estimates that 7 to 13 percent of all preschool children snore.[40] It was at one time thought to be harmless, but now research shows that "primary snoring" (snoring in individuals without OSA) in childhood causes problems in attention and memory,[41] greatly increases the chances of having an epileptic seizure,[42] and can be a warning sign of developing more serious problems.[43] Scientists have identified chronic snoring as a likely indicator of sleep apnea.[44] Unfortunately, many physicians often blame childhood snoring on tonsils, adenoids, and allergies, missing out on one likely root cause—poor oral-facial health.[45]

As a result of mouth breathing[46] and the chronic snoring that may accompany it, OSA is becoming more common among young children.[47] Sleep specialist Dr. James O'Brien has said, "Anyone who snores will develop obstructive sleep apnea, if they live long enough."[48] As airway-centric orthodontist (one heavily concerned with not restricting the breathing tube) Bill Hang put it subsequently, "Nowadays kids are taking on what was once mainly a burden for adults."[49]

Disturbed breathing during sleep has also been linked in some studies to ADHD (attention deficit hyperactivity disorder) in children, characterized by overactivity, trouble focusing, and/or difficulty controlling their other behavior.[50] Stephen Sheldon, sleep specialist at Lourie's Children's Hospital in Chicago and author of *Principles and Practice of Pediatric Sleep Medicine,* estimated that 75 percent of children with ADHD could trace their problem to sleep-disordered breathing.[51] Abnormal patterns in breathing during sleep may affect a child's heart as well as the cardiovascular health of older persons. Sleep physician Christian Guilleminault and cardiologist John Schroeder believe that "if we do not address the OSA problem in children we will not be able to reverse the cardiovascular problems that will occur."[52] Those problems appear to include cardiac arrhythmias and high blood pressure.

Besides the suggested sleep–ADHD and heart disease connection,[53] some studies link childhood problems like nightly bedwetting (enuresis)[54]

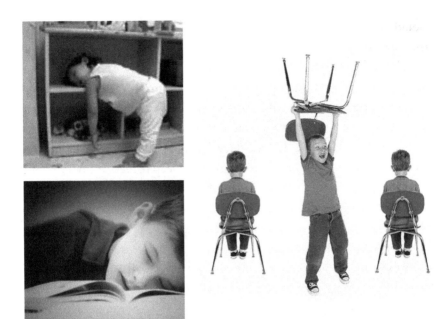

Image 35. Sleep breathing disorder (SBD) symptoms in children may include snoring, pauses in breathing, restlessness, or abnormal sleeping positions (left). Behavioral problems such as attention deficit hyperactivity disorder (ADHD) may result (right). Note open mouth.

and grinding of teeth (bruxism) (see Box 1), to sleep-disordered breathing.[55] Of course these studies only show correlations; the causes of each may be more complex.

More difficult to evaluate than the direct impacts of OSA is the large array of stress-related diseases that are partly traceable to a history of OSA and disturbed sleep.[56] The key fact is that mouth breathing can lead to OSA and interrupted sleep, a source of stress.[57] This stress, as we have seen, may contribute later in life to high blood pressure and heart disease;[58] it also may create problems with vision,[59] chronic lung disease (COPD, chronic obstructive pulmonary disease), allergies, cancer,[60] Alzheimer's disease, and other ailments.[61] Recent research shows that sleep apnea may also compromise the blood–brain barrier. That barrier generally keeps harmful bacteria, infections, and toxic chemicals from entering the brain. A weakened blood–brain barrier is associated with significant brain damage in conditions such as Alzheimer's disease, stroke, epilepsy, meningitis, and multiple sclerosis, among others.[62] We must note, however, that these associations should not be overinterpreted—much more research is required to establish the existence of any causal relations.

OSA unfortunately often develops without a victim or a parent being aware of it and can affect a person's well-being from an early age. For instance, sleep-disordered breathing can significantly degrade a child's mental functioning. This is not just a consequence of increased sleepiness; sleep scientists Dean Beebe and David Gozal have developed a model for the likely mechanism.[63] In other words, disturbed sleep reduces the brain's ability to think and control behavior, and different deficits are related to different degrees of apnea.[64] Interestingly, there is some evidence to suggest that high intelligence is protective against the cognitive problems seen widely in those with sleep apnea, putatively (and in our view doubtfully!) because they have more "reserves" in brain power.[65] Gozal and his colleagues also estimate 2 to 3 percent of preschoolers may already be suffering from OSA.[66] Snoring and behavioral problems, as we've seen, can be early warning signs of the development of OSA and other dangerous breathing dis-

---

BOX 1: CLENCHING AND GRINDING THE TEETH

Dentists are deadly enemies of teeth clenching, gnashing, and grinding—"bruxism"—activities that are not directly connected to eating or talking. And why shouldn't they be? After all, bruxism wears down the teeth. People who grind and clench their teeth usually have unevenly erupting teeth where some make contact with those of the opposite jaw well before others. These are referred to as "premature contacts." There can be many reasons, as we have seen, for such unfortunate jaw development, but individuals who grind their teeth don't have the upper and lower teeth resting in light contact for at least 8 hours a day.

Grinding and clenching are reactions of the body to premature contacts that keep erupting teeth from fitting properly. If proper oral posture is maintained at least 8 hours a day, with the teeth lightly in contact, the teeth of the upper and lower

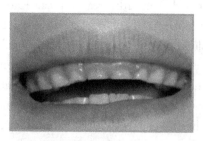

Image 36. Grinding of teeth as the child attempts to open her airway can produce obvious wear and makes noise that also can alert parents.

orders.[67] Because they are viewed both as common in industrial societies and not recognized as related to sleep and breathing, these symptoms are often ignored. Fortunately, parents are beginning to take their children to health care providers to evaluate their sleep—the "how" not just the "how much"; that is, the *quality* of their children's rest.

Image 37. Disheveled covers and an open mouth during sleep are signs of trouble ahead.

This is the place for an important caveat. *When considering the ills to which mouth breathing and OSA "may contribute" remember that does not mean "does contribute"; "is associated with" is not the same as "causes."* It took decades of investigation to establish firmly the link between smoking and lung cancer, a relatively simple task because it's usually easy to determine how

jaws will maintain their relationship so that all meet evenly and bruxism is avoided. Bruxism is often blamed on stress, which may indeed be involved but not as the original cause. Dentists, who do not understand this posture concept, teach patients to keep the teeth out of contact to avoid damaging "precious" enamel; therefore the problem is never cured because grinding is required to cure it—the cure is more of the same. By continuing to keep the teeth apart, the problem just continues to get worse. Dentists may prescribe use of an acrylic orthotic plate that keeps teeth apart and protects enamel, so the body never finds an equilibrium in which the teeth are in even contact. It is Sandra's opinion that if patients have the premature contacts filed down and are taught (most of the time a dental appliance is needed) to keep their teeth in light contact for a third of the time, the grinding problem would disappear.[i] That, of course, doesn't mean that more fundamental problems, like restricted airway space, will also vanish.

For some inexplicable reason the concept of keeping the teeth apart (freeway space) at rest is taught at every dental program in the world. Here lies another aberration of the dental practice that we would like to challenge using common sense.

i Based on a discussion between Dr. Antonio Facal Garcia and Sandra, 2016.

many cigarettes per day and for how long people smoked—a lot easier than finding out, say, whether one's mother had OSA during pregnancy. Furthermore, some heavy smokers never get lung cancer, and some people who have never smoked die of that disease. Our goal is to make you aware of the many *possible* consequences of poor oral-facial health, not to convince you that a child who is a mouth breather is condemned to a short life of misery or that correcting its jaw development will guarantee a long, healthy, happy life.

Treating Obstructive Sleep Apnea

Although our focus is on children here, we always must remember that children's oral-facial health issues, if not appropriately treated, likely will follow them into adulthood, sometimes with miserable consequences. One of the main long-term symptoms of OSA is perpetual sleepiness, which can take a severe toll on a person. Paul himself became aware of this when a friend showed telltale symptoms and was diagnosed with sleep apnea. Along with her sleep apnea came exhaustion; she was so tired during the day that she had difficulty functioning. She was unable to work, which put a great deal of stress on her and on her family. Seeking relief, she went through a very painful operation but unfortunately was not cured.

That case showed what a dreadful and costly disease sleep apnea can be. A different operation, maxillomandibular advancement surgery (MMA), which involves moving the upper and lower jaws forward, is the only real alternative for curing adults with particularly serious cases of OSA.[68] In one recent case MMA surgery was shown to correct increased pressure within the skull (intracranial hypertension, IH) of a 44-year-old woman with OSA who suffered from headaches and then a sudden onset of "brain fog." The operation resulted in great improvement in the symptoms of both OSA and IH.[69] Sadly, there are very few surgeons who can competently perform that type of operation and still fewer maxillofacial surgeons who can virtually guarantee improvements in OSA.

Whether the ends justify the means of various jaw surgeries for OSA, considering the pain, long recovery periods, and risk of infection and even death, remains controversial,[70] especially because surgery can in some cases worsen airway problems.[71] Even highly trained and respected surgeons shy away from this procedure. William Bell, known as the "godfather of corrective jaw surgery," described it as "too complicated, too invasive, too time-consuming, too expensive and too unpredictable."[72] The most common surgery for airway problems does not deal with the jaws themselves. That is the removal of tonsils and adenoids, which tends to relieve sleep apnea temporarily in kids but has limited results in adults. There are many other options, some better than others.[73] For example, uvulopalatopharyngeoplasty (just say "UPPP"—"U-triple P") and other soft tissue surgeries like tongue reduction continue to be practiced on a limited basis.

Surgery is a course of last resort, which is a good thing, considering the many millions of people who suffer from OSA. Ultimately, the only true solution to the OSA problem is prevention from an early age. But for those who have already developed the disorder, there are fortunately alternatives that can give some relief in all but the most severe cases.

Many people can gain relief by using continuous positive airway pressure (CPAP) machines, which often can eliminate some of the worst symptoms of OSA and in some instances offer near complete relief. A personal anecdote: A CPAP machine was enormously helpful to Paul's grown daughter. Those, like her, who can use them in comfort often report great relief. Sadly, though, some children now are forced to deal with the complex apparatus at an early age.

These machines force air into the nose at pressures high enough to reopen obstructed airways. They cost several hundred dollars and, at least in the United States, require a doctor's prescription, so those who are less well off, not medically connected, or lack insurance may not be able to afford the machines. Moreover, many people find it difficult to sleep wearing a mask and dealing with the tubes connecting them to the machine, although design improvements in recent years have helped with this problem. Furthermore, CPAP

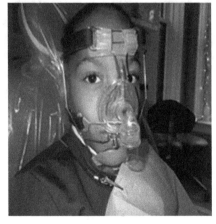

Image 38. Use of a Continuous Positive Airway Pressure (CPAP) machine can help alleviate some of the worst symptoms of Obstructive Sleep Apnea (OSA). "You can see this is no way for a child to live." (photo courtesy of Dr. Kevin Boyd and Dr. Steven Sheldon)

machines will not always prevent the disease from worsening—depending on causative factors—and additional approaches to treatment may be useful. With postural treatment and reduction of obesity, some adults can greatly improve their airway functioning, and in some this can lead to improvement in cognitive functioning.[74] The connection between obesity and OSA is a result of an increasing neck circumference associated with weight gain.[75] Large fat deposits around the airway put pressure on the pharynx, narrowing it in some cases to dangerous levels.

The only truly effective way to cope with the OSA epidemic is to recognize its seriousness and to treat its potential development in the first decade of life through postural training if needed and, sometimes, corrective action. Fortunately, a program directing growth of the jaw, face, and airway called "orthotropics" and more recently by us "forwardontics," the subject of the following chapters, promotes development of the face to prevent

---

BOX 2: FOLLOWING THE BULLDOG TRAIL?

A leading dentist and a founding member of the American Academy of Physiological Medicine and Dentistry (AAPMD), Mike Gelb, has compared respiratory trends in bulldogs, which are notorious for breathing problems, to trends in sleep apnea in humans. The bulldog has what has become known as the short-headed ("brachycephalic") syndrome. Forward growth of bulldogs' faces is restricted, their nostrils are narrowed, their teeth are crooked, with upper teeth biting behind the lower ones (crossbite), an elongated soft palate partially blocks the airway, and their tongues can exceed their jaw capacities. Like human beings with some of these problems, they tend to suffer the difficulties of obstructive sleep apnea, and, as with people, it is worsened by obesity.

This dangerous pattern is a result of long-term selective breeding for extreme characteristics that affect the dog's health. Human culture here is negatively influencing the dog's genetic endowment. In people themselves human cultural change (industrialization) has caused developmental changes that parallel those caused by changes in the bulldogs' genes. As a result of the latter, bulldogs may be headed for extinction. Here is a series of pictures of different the University of Georgia bulldog mascots from 1956 through 2011. The first one died when she was eight years old, and the last one died when

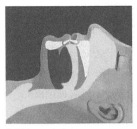

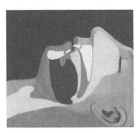

(a) Normal sleep   (b) Snoring   (c) OSA

Image 39. Muscles that control the tongue and soft palate hold the airway open during normal sleep (a). When these muscles relax, if there is not enough space with forward placed jaws and good muscle tone, the tongue falls back and the airway narrows, leading to snoring (b). If obstruction is severe, and as we lose muscle tone with age, alcohol, or obesity, the airway may collapse and become blocked, which obstructs breathing (c).

she was two years old. Their life-support systems, from the autonomic nervous system to the cardiovascular system, were compromised because their airways were too narrow, as are those of the increasing number of human beings. Our breed is not threatened with extinction by sleep apnea or lack of oxygen, but it seems clear that the well-being of millions of people is. Some bulldog fanciers have seen the errors of the past and taken steps to alter the evolutionary course of the breed. We can't do that in humans, but we can encourage practices of chewing and good oral posture in our children.

Image 40. Series of pictures of different bulldog mascots in the University of Georgia from 1956 through 2011, showing the evolution of the breed. The first one died when she was eight years old, and the last one died when she was two years old. These dogs have breathing issues and short life spans. This may be an example of where humans are headed.

dental crowding, allows the mouth to function optimally, and averts sleep-disordered breathing. Unfortunately, as we'll explain, following a forwardontic program at the moment usually takes a long time, dedication on the part of a patient, and competent professional assistance, which may be hard to find. That's one reason why we emphasize the importance of prevention.

CHAPTER 7

WHAT CAN YOU DO?

If our discussion so far of a widespread oral-facial health epidemic has gotten you concerned about it, the natural question to ask is, "What, if anything, can I do?" How can I prevent oral-facial health problems from arising or, if they've already begun, prevent them from getting worse? In thinking what overall answer to give, we were tempted to suggest simply following George Catlin's 19th-century advice to "close your mouth and save your life."[1] Breathing through the nose with mouth closed, with teeth in light contact and the tongue resting on the roof of the mouth, remains key to preventing, or at least ameliorating, the ills we've discussed in previous chapters. But of course that's just the beginning. What follows are suggestions, based on our interpretation of the literature and the extensive clinical experience of Sandra and her colleagues. The points are mostly aimed at what you should do with your children, but some will be useful for you as well.

Breastfeeding and Baby-Led Weaning

As we discussed in Chapter 5, breastfeeding for as long as possible is associated with reduced malocclusion.[2] At present the recommendation is that exclusive breastfeeding should last about six months.[3] Equally important is how babies are weaned. Weaning should not be a single act, but, for proper oral-facial development, ideally it should also stretch over a substantial time. Wean with solid foods that will keep the jaws moving as much as possible, taking care to watch for choking. Avoid the pap called "baby foods," almost all of which are soft and sweet,[4] and other soft processed foods.[5] Overall, it seems best to prolong breastfeeding after the introduction of alternate food,

Age 8

Age 15

Age 7

Age 11

Image 41. Sisters (with similar genes) who were instructed to close the mouth; one succeeded (bottom) and the other one did not (top). Neither of them had any treatment. Note the obvious difference in their teenage years in facial development and attractiveness. (Courtesy of Dr. John Mew.)

Image 42. Weaning into solids foods can give kids a head start into proper jaw development.

Image 43. Babies are capable of feeding themselves proper food; in other words—no more mush!

and the kids should be started as early as possible on alternatives that are solid and not sweet.

One method to consider is "baby-led weaning" (BLW),[6] which simply means letting your child feed himself from the beginning of the weaning. The term was used originally by Gill Rapeli, midwife and health consultant. BLW is an alternative approach for the introduction of complementary foods for babies in which the child feeds himself with his hand instead of being fed with a small spoon by an adult. The idea is that through the BLW approach, the baby and his parents will share the family food, and when it's time to eat the mother continues to offer milk (the ideal is for the milk to be breast milk) until the child auto-weans.

Chewing

Chewing also influences how the face is formed, so as children begin to eat solid foods, parents should teach them good chewing habits. Of course, it means spending time with the kids. For example, Sandra's two early-teen-age kids take the opportunity at mealtimes to sit with parents and grandparents and rehash the events of the day. With working parents and numerous children's after-school activities, it is difficult to maintain such family traditions, though. A way must be found to reach a happy medium, where dinner can be a social occasion, during which good oral posture can be fostered. Sandra and David encourage their children to both talk and eat at the table but to slow down and not do both at the same time. Paul and Anne's kid, Lisa, is now difficult to train, as she's a grandmother.

Posture

As we outlined in Chapters 5 and 6, proper oral—and perhaps overall—posture is crucial to healthy oral-facial development. The ideal oral-facial resting position entails three things: lips closed, tongue on the palate, and teeth touching lightly together. Teach your children when not eating or talking to rest with their mouth closed. Start immediately after birth. Close your baby's lips lightly with your fingers for a few seconds when she finishes nursing. At the very worst it's a harmless action, and future studies may show it is of considerable benefit. Catlin believed it helped generate excellent oral posture, good health, and fine appearance in Native Americans not in close contact with Europeans.

Dr. John Mew, the founding father of orthotropics, covers the posture-related issues very well. In a letter he sent us, entitled *Old Fashioned Rules Create Good-Looking Faces,* he wrote:[7]

Surprisingly Great-grandmother's advice was often right and some simple remedies can be very effective. The face is very sensitive to oral habits when a child is young, and simple things like leaving the mouth open too much can make a big difference. It used to be considered important to keep your lips closed especially while eating but many sociologists these days think it wrong to control children too strictly.

In Victorian times children were expected to keep quiet until spoken to and always be polite to older people. Some parents consider this as restrictive and allow their children to play with their iPads many hours a day, often with their mouth open, their neck bent and their lips apart. This really can do a lot of harm to the way the face develops.[8]

As children leave infancy they need regular training to lead a healthy life, which means a lot of focus on posture. Without the necessary exercise of chewing, as we saw in Chapter 2, the human jaw does not develop the muscular capacity to keep itself closed. However, jaw muscles will not do the job by themselves; children must learn good habits and muscle memory. Obviously, a child cannot constantly be thinking about having his mouth in the proper posture, so muscle memory must be established to activate the muscles in the jaw even while the child is distracted or sleeping. These periods of rest are where the "posture" part comes in.

Some practitioners claim the loss of proper posture for the entire body affects our oral posture, and vice versa. That makes intuitive sense, but more research would be very useful as there is currently little scientific evidence to tie poor overall posture to the oral-health epidemic. Here we are leaning mostly on anecdotal evidence from posture specialists like Esther Gokhale to speculate that improving overall posture might help with oral posture. That it can help with such things as back pain can be attested to by many of us who spend too much time slouched over computers!

As Jared Diamond points out in his pioneering book *The World until Yesterday,*[9] the environment a child is exposed to will determine her muscular, or postural, reaction. In many indigenous settings, children develop

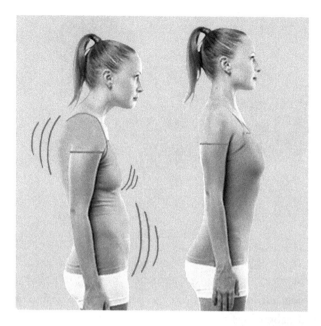

Image 44. Connection between overall posture and oral posture. The way the head is held influences the pressures on the jaws during growth and in turn is influenced by body posture.

appropriate body posture because of the way in which they are brought up. There is only limited information on the relationship between oral posture and overall posture,[10] but it seems likely that head-forward posture is related to malocclusion.[11] A natural posture for a baby in a carrier should be one that many preindustrial societies still foster: hips thrust back and back straight, vertical and able to look in the same direction as the caretaker—starting to build toward correct adult posture. Carrying devices common among indigenous peoples tend to align the hips with the spine, promoting good overall posture and thus oral posture. This doesn't mean

Image 45. Woman from traditional culture nursing her baby. Note her straight back, toned yet relaxed at the same time. She is encouraging proper development of her baby's hips and back by nursing it at a natural angle, which simultaneously adds to bonding.

Image 46. A modern girl, right, can imitate the beautiful healthy posture of the more traditional girl with her straight back, closed lips, and relaxed shoulders.

that you should tie a sheet around your neck and put your child in it; there are some more technologically advanced options. Ergonomically designed strollers, car seats, and baby slings are available on the market, and these may serve as adequate substitutes.

Image 47. Natural posture for a baby should be that fostered by many preindustrial societies: hips back and back straight, vertical, and looking in the same direction as the caretaker—starting to build toward correct adult posture.

Humans evolved to be carried as babies and have close contact and interaction with their mother.[12] Being able to see the mother's face is critical, for example, for language development; usually orienting the baby away from the mother at such a young age would hinder his development.[13] However, even though facing the mother provides the baby with important opportunities for interaction,[14] it presents views very different from the ones seen by the mother. As the baby matures, his cradling position typically shifts from the chest to the back, enabling the mother to cope with the increasing weight of her child and enabling the infant to see the world as his mother sees it and, quite likely, to develop good overall (and oral) posture.

The bottom line here is that if we do not help children build oral strength and learn to use correct posture, their jaws are unlikely to develop well. Holding the jaw with all teeth in contact at rest was in the distant past, we speculate, a natural result of such things as late weaning, tough foods, and fewer stuffy noses. Now it is a skill that should be taught until it becomes part of one's implicit memory. And, of course, you should continue to feed your children harder, minimally processed foods and encourage them to chew thoroughly. At the same time, adopt and foster good table manners relevant to oral-facial health, especially that to keep the mouth closed while chewing. Teach your children to eat slowly, pace themselves, and include lots of pauses with lips closed. Encourage them to speak slowly and understandably, all, as we will see, part of their postural training.

BOX 3: "CULTURAL CHANGES TO REST AND LEISURE"

Working constantly is a cultural condition; it was not our ancestors' perpetual natural state. Our ancestral hunter-gatherers typically had a surprising amount of leisure time because their main concerns were to provide food and shelter for themselves and their families; they worked only as much as required to feed and protect themselves[i] It is often thought that indigenous peoples spent their entire day hunting and gathering for food. However, the truth is quite the opposite.[ii] Writing of the !Kung bushpeople, anthropologist Richard Lee commented:

A woman gathers on one day enough food to feed her family for three days, and spends the rest of her time resting in camp, doing embroidery, visiting other camps, or entertaining visitors from other camps. For each day at home, kitchen related routines, such as cooking, grinding nuts/herbs, collecting firewood, and fetching water occupy 2 to 3 hours of her time. This rhythm of steady work and leisure is maintained throughout the year. The hunters tend to work more frequently than the women, but their schedule is uneven. It was not unusual for a man to hunt avidly for a week and then do no hunting at all for two or three weeks. Since

(continued)

i M. Sahlins. 1972. *Stone age economics*. Aldine.
ii J. Gowdy. 1997. *Limited wants, unlimited means: A reader on hunter-gatherer economics and the environment*. Island Press.

hunting is an unpredictable business and subject to extremely variable climatic control, hunters sometimes were forced to stop for upwards of 3 weeks. During these periods, visiting, entertaining, and especially dancing are the primary activities of men.[iii]

Of course, few of us can adopt a !Kung-type lifestyle or other traditional one. But by focusing more on oral-facial health we can gain more of the badly needed uninterrupted sleep that often seems so scarce in industrialized societies.

Industrial societies by comparison now force most everyone to work more avidly and for longer hours. We do not go to sleep when the sun sets; we turn on the lights so we can keep going, though it may not always be for work. A recent study, for example, concluded that a third of Americans got an inadequate amount of sleep, less than 6 to 8 hours.[iv] The toll this takes on our body is often blatant, especially when the impacts of OSA are added in, but it has become so common that we consider it normal. Having bags under your eyes or falling asleep during the daytime would have been considered a complete abnormality in a preindustrial society. Not so today.

Are our lives worse because artificial lighting allows us to work much longer into the night, with potential physical and mental repercussions? Was it better back in the Stone Age when we breathed clean air, drank clean water, exercised regularly, but died at an average age of 40? Human evolution has had an insufficient amount of time to adjust genetically to the dramatically

iii R, B, Lee. 1969. !Kung bushmen subsistence: An input-output analysis. Contributions ronntizropol-ogy: Ecoiogicoi essays. *Natural Museums of Canada Bulletin.* 230: 73–94.
iv Morgan Manella. 2017. Study: A third of U.S. adults don't get enough sleep. CNN. Available at http://cnn.it/1QUV07R.

Image 48. Being bent over our phones is not the problem; it's our posture when we do it. These women are bent many hours a day, but look at their straight backs. Coincidentally they also have well-developed jaws and straight teeth.

Image 49. In the days of yore, proper people sat up straight and also had good oral posture. They kept their mouth closed and their teeth were straight.

changed environment in which we operate. Is it ever likely to, since most of the issues raised are ones that seem to affect people over reproductive age, especially among women? Whether the scale of problems such as suffering with poorly developed jaws and sleep apnea are actually counterbalanced by the advantages of artificial electric lighting[v] and living in tight, weatherproof shelters, are not all that clear. What is clear is that fairly simple steps could help reduce some of the negative effects of industrialized environments.

Obesity is often considered an effect of inactivity. It may often be a cause as well, in a proverbial "vicious circle," where one condition feeds the other. Obesity is a result of poor diet and lack of exercise, but it can also be a postural sign that a child, or for that matter an adult, does not have enough energy to be sufficiently active during the day. Obesity promotes poor oral posture, which may lead to sleep apnea and, in turn, a deficiency in nocturnal rest.

Obesity aside, many children are relatively inactive in part because they don't have the energy to be physically active much of the time. Why don't they have sufficient energy? In many cases, it may be because they haven't been sleeping properly, in quantity or quality. For their health and development, many children need to slumber more, and better, than they do.

Not only for our well-being, but also for the well-being of the children we bring up, we need to rethink the balance of activity to rest in industrial societies. We have replaced practices that have been proven through tens of thousands of years of evolution, simply because it was easier, not because it was better. For perhaps the first time ever, society may be forced to reverse its actions, to take a step back in the timeline. Without the proper environment, children will be negatively impacted, and their improper and/or insufficient sleep may bring deformity, disease, and despair.[vi]

v K. J. Navara and R. J. Nelson. 2007. The dark side of light at night: Physiological, epidemiological, and ecological consequences. *Journal of Pineal Research* 43: 215–224.
vi J. S. Durmer and D. F. Dinges. 2005. Neurocognitive consequences of sleep deprivation.. *Seminars in Neurology*: 117–129; J. M. Mullington, M. Haack, M. Toth, J. M. Serrador, and H. K. Meier-Ewert. 2009. Cardiovascular, inflammatory, and metabolic consequences of sleep deprivation. *Progress in Cardiovascular Diseases* 51: 294–302.

Sleeping

Parents should also keep an eye on children's sleep. If they're routinely tired in the morning, one of two things is likely occurring: either bedtime is too late at night or early symptoms of sleep disorder breathing are manifesting. Even if you don't notice any nocturnal activity disturbance (snoring, tossing and turning, and the like), a child's chronic lack of energy should make you suspicious that she is not resting and sleeping properly.

Remember that airway distortions tracing to poor oral-facial development can be a great enemy of adequate rest.

Breathing and Allergies

Aside from the occasional cold, a stuffy nose in a young child is no joke. It's important to pay attention to allergies and breathing from the beginning. The very first problem for oral posture can occur within hours of birth. Tiny babies can get stuffy noses easily. Pay attention to the environment where they rest. It is important to protect kids from common allergens and other particles that collect indoors.

Imagine you go up to the second floor in your house on a sunny day. There is a ray of sunlight crossing the room. What do you see? Billions of tiny particles! The nose is designed to filter them out. It functions as a "scrubbing tower"; it traps many airborne particles before they can reach the lungs.[15] But if the nose is stuffed up and you breathe through your mouth, many more particles will enter your lungs. Scientists know that inhaling these particles can be detrimental to your health, although much remains to be done in determining which ones cause which effects. This is not surprising; assigning blame is often close to impossible, even with intensive study. As air pollution expert Professor Kirk Smith has commented, "After tens of billions of dollars, tens of thousands of studies, and now about 70 years of intensive effort, scientists still do not know what it is in tobacco smoke that causes the health effects observed."[16] Minimizing inhalation of particles as well as doing what you can to keep the nose at peak performance should be a priority if you are concerned with the oral-facial health epidemic and its consequences.[17]

Unhappily, there are a huge number of common indoor air pollutants[18] as well as outdoor ones. But the concentrations of pollutants indoors often exceed those outside. A case in point is formaldehyde, a toxin emitted by some kinds of furniture or construction that can worsen the upper respiratory symptoms of children sensitive to other allergens.[19] One good rule of thumb is to avoid needless aerosol sprays and volatile products where young children are living. Controlling cockroaches and mold is also important, as is frequently bathing cats, dogs, and other household pets. Increasing ventilation can also help, and one or another of various air cleaning devices could be considered, especially during the first years of a child's life, although their effects are debatable.[20]

One place where very young children easily acquired stuffy noses is in day care centers. Colds can be passed around by caregivers, who help youngsters wipe and blow their noses but sometimes do not practice good sanitation. Simple training in hand washing has been shown to significantly reduce colds among the children under 2 years old,[21] an important time for jaw development.

Many parents complain that because their kids have allergies, they can't breathe through their nose. But consider the possibility that in some cases mouth breathing, taking many more particles into the lungs than would occur through nose breathing, may actually be the root of the problem. Christian Guilleminault, for example, an expert on sleep apnea, has determined that enlarged tonsils and adenoids are more likely a result of mouth breathing than a cause of it.[22] Respiratory allergic reactions may follow a similar pattern: it may not be always the stuffy nose that makes kids breathe through their mouth; it may be mouth breathing that invites in allergens, which in turn leads to stuffy noses, and the cycle continues with further mouth breathing as a result. Adding to the complexity of the situation is the considerable individual-to-individual variation in susceptibility to allergens, and the time-to-time variability in their presence (think pollens).

In addition to trying to limit exposure to allergens and pursuing medical help controlling allergies, we recommend looking into alternative therapies that might help, such as the Buteyko breathing technique, from which both older children and adults can benefit.

The Buteyko Breathing Technique

The perfect man breathes as if he didn't.
Lao Tse, philosopher. 6th century BCE, China.

The Buteyko method is designed to help anyone, child or adult, train themselves to nose breathe and increase their respiratory efficiency.[23] We are not Buteyko therapists, but in Sandra's clinical experience some patients have found this method extremely helpful in reducing mouth breathing; other clinicians have reported similar experiences with the use of this method.[24]

The physician Konstantin Buteyko (1923–2003) originally developed the method that bears his name in the Soviet Union around 1950. The success claimed for the method in the treatment of asthma is controversial, but we have found that it does help in making a transition from habitual oral breathing into exclusively nasal breathing. The Buteyko method is based on standard medical principles related to the processes that deliver oxygen to your cells.

According to Buteyko, normal breathing:

- Should not be seen.
- Should not be heard.
- Should involve a closed mouth (with the lips in slight contact).
- If breath is seen, heard, or done with the mouth open, it is overbreathing.

Buteyko exercises are designed to help a person achieve a conscious reduction in breathing's frequency as well as its volume. The method can be thought of as "breath retraining": the aim is for the new breathing pattern to take hold through repeated practice, as an instance of implicit learning, as in learning to ride a bike, so that it becomes "second nature."

Keeping the nose unclogged and promoting nasal breathing during the day will improve night sleep. Restricting ourselves to nasal breathing while exercising is another key element of the Buteyko method. Improving sports performance can be an added benefit of using the method, which could help motivate your child to avoid mouth breathing. Optimizing oxygen consumption is vital for a good sport performance, and that's what the Buteyko method attempts to do by overcoming the hyperventilation syndrome that reduces the carbon dioxide concentration in the blood and impedes the blood's ability to deliver oxygen to tissues.[25] The nose is designed to avoid dehydration of the lungs by humidifying incoming air,[26] which has helped human beings to thrive in a wide variety of climates.[27]

There are many stories about the importance of nose breathing to long-distance runners. For instance, the Tarahumara Native Americans, claimed by some to be the greatest marathoners in the world, run up to 60 miles a day, minimally shod, sometimes barefoot, in the Copper Canyon of northern Mexico. They breathe almost exclusively through their noses, and maintain unstressed, peaceful faces. And Apache "spirit runners" have been trained from childhood by running in the desert while holding a mouth full of water. They learned to breathe deeply and rhythmically through their noses, avoiding the panting that would dry their throats in the dry desert air. Shades of George Catlin's Native Americans!

The three Buteyko activities that Sandra believes are particularly effective in promoting nose breathing are sleeping with a tape over the mouth, counting steps, and unclogging the nose.

Image 50. *High-performance nose-breathing technology.* Runners Patrick Feeney and Chris Giesting used high-performance nose-breathing technologyas they paced their four-man team in the 4 x 400 meter relays for Team USA over two days to win the gold medal in the IAAF World Indoor Championships and nearly break the world record. When Feeney was asked about high-performance nose-breathing technology stated, "After a couple of weeks working with high-performance breathing technology's oxygen advantage program, I have been sleeping better than ever and feeling much more calm and relaxed. It has helped me focus on the race at hand, trust my training, and get my mind right for the best times I have ever run."

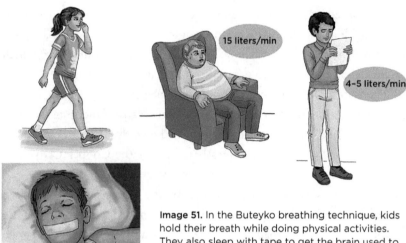

Image 51. In the Buteyko breathing technique, kids hold their breath while doing physical activities. They also sleep with tape to get the brain used to breathing exclusively through the nose. A fit person will have better respiratory efficiency, taking in fewer liters of air per minute.

Tape

Patients without any other severe problem (such as a deviated septum or severe allergies) who mouth breathe are encouraged by Buteyko therapists to place a piece of hypoallergenic tape or Micropore over their lips at bed time. It's harmless and easily broken through or removed, but in the meantime the tape serves as a reminder when the impulse to mouth breathe arises.

This practice may seem shocking for some parents and children, but we believe use by older children or adults will be both safe and can be beneficial. *Needless to say, one should not tape shut the mouths of infants or small children!*

Sandra herself found the Buteyko method very useful in her family. Her 89-year-old father, her husband, and her teenage son used it with enthusiasm. All of them claimed to be sleeping better as a result, waking up fresher and with no dryness in the throat. Her father, who suffered from COPD (chronic obstructive pulmonary disease, including what used to be called emphysema and chronic bronchitis), reported a huge spike in the quality of his life when he started to practice the Buteyko method: it seemed to reduce the anxiety and overbreathing that accompanied his inability to inhale enough oxygen.

Counting steps

The count of steps is an exercise in which patients pinch their nose and walk while someone counts the number of steps they can take without breathing through either nose or mouth. This activity is usually done under the supervision of a certified Buteyko therapist. With this practice of breath control, the number of steps will increase as the patient becomes fitter. There are now apps for smartphones that count the steps, making practice at home easy.

Unclogging the nose

Here's the way to do unclog your nose, according to Buteyko specialist Patrick McKeown.[28] Try it yourself the next time you have nasal congestion:

- Sit down.
- Take a small breath in through your nose.
- This breath should make no noise.
- Breathe out through your nose.
- Then pinch your nose with your fingers so that the air cannot come in or go out.
- Gently nod your head up and down.
- Do this for as long as you can.
- When you need to breathe in, then breathe in through your nose only and try not to let the air sneak in through your mouth.
- Calm your breathing as quickly as possible.
- Wait about half a minute and practice this again. Your nose will be unblocked by the third attempt. If it is not; practice this again until your nose is unblocked.
- If your nose becomes blocked again; practice the exercise again.

Good Oral Posture Exercises (GOPex)

Sandra uses a good oral posture exercise program (called GOPex) in her practice. It is a set of simple exercises developed by dentist Simon Wong, a pioneer in finding solutions to oral-facial problems. Through this program, kids—and adults—learn to be still by *doing*, because function and posture are intertwined. GOPex is a type of "myopostural" therapy ("myo" for muscular), not to be confused with oral-facial myo*functional* therapy, which targets function—movement—but plays a minor role in oral-facial growth and development.[29] GOPex is designed to develop correct oral posture and to achieve balanced growth of faces, throat, and teeth.

The exercises are partly based on good old-fashioned table manners: sit

Body Posture
Checklist

- Crown of head uplifted, as if it were gently lifted by a helium balloon;
- Mouth completely shut;
- Shoulders rolled back;
- Belly tucked in, under ribcage;
- Hips forward and butt back;
- Hands relaxed on the top of your thighs;
- Knees bent about 90 degrees in a chair that fits you properly;
- Both feet apart on the ground, at the same distance as your shoulders.

lips
chin
neck
shoulders back
hands
hips
knees
feet

Image 52. In GOPex (good oral posture exercises). body posture and oral posture depend on each other.

up straight, keep your mouth closed, don't chew with your mouth open, don't swallow your food half chewed, and the like—which turn out be remarkably good advice for oral-facial health. The exercise routine also promotes slowing down when eating or speaking; emphasizing pauses is central to learning the correct resting position. By learning to adopt the ideal rest position in between bursts of mouth activity (chewing or speaking), the brain learns the default position through repetition, and the proper rest posture will come to be held unconsciously through much of the day. Which children should be encouraged to put in the effort the GOPex program requires is one obvious question. Sandra's answer is only those who have signs of developing problems, although she adds that those signs can be easily missed by the untrained eye of a parent. So careful attention to the sorts of symptoms we describe is recommended.

The GOPex exercises train children to:

- Chew as if they mean it
- Swallow with their teeth together
- Close their mouths when not in use
- Use their noses for breathing

Here are the essential GOPex instructions designed to help children find and maintain correct oral posture.[30]

First, an exercise to build good jaw-muscle tone:

Meaningful Chewing. Reserve 2 to 3 minutes for at least one meal a day to focus fully on chewing. Always chew your food until it liquefies. This will aid digestion and, more importantly, the amount of chewing required will help build sufficient muscle tone to maintain a good closed-mouth posture.

With each mouthful of soft food, try to build up to a minimum of chewing 15 times before swallowing, and try 20 chews for hard food. As your muscles get stronger, you may not need as much chewing before your food liquefies.

Always chew with lips together and always begin your swallow with teeth of upper and lower jaws touching. Consciously pause for at least a 2 seconds before starting the swallow. Focus on the pause.

Second, two exercises to encourage nasal-only breathing:

Counting Exercise: Counting out loud slowly from 1 to 60 (30 for the very young). Pause between every number and touch your teeth together and your lips together only once. After each five counts, pause to breathe through your nose. Repeat this exercise at least once every morning and once every evening. All in-breaths should be through the nose and only on every count of 5; out-breaths can be through your nose or naturally through your mouth as you count out loud.

Reading-Out-Loud Exercise: Speaking with good punctuation is an excellent practical application of nasal-only breathing. Take 5 to 20 minutes each day to read aloud, pausing at each comma and full stop in the sentence to close your mouth and breathe "in" only through your nose.

Third, we have suggestions to make good oral posture your natural state. Incorporate the exercises just described into your everyday life. Here are some suggestions on how to achieve this:

- To further develop controlled breathing and increase your stamina, concentrate on nose breathing when exercising. When you speak, pause regularly, inserting "punctuation" between ideas, and inhale only through your nose. Take time each day (perhaps with another family member) to do some conversational exercise with the goals of keeping your mouth closed when not speaking and breathing exclusively through your nose as much as feasible.

- Start with walking and keeping your mouth gently but completely sealed. Try to keep this up a little longer each time until you can comfortably maintain this for 5 minutes, then go longer. Eventually try this while running or jogging, if that is part of your routine exercise. Over time your body will become more efficient in its oxygen exchange,[31] and your stamina will improve.

- Stand up straight in front of a mirror, open your mouth and smile with teeth showing. Check yourself to make sure the corners of your mouth go up evenly and that you feel relaxed. Practice this a good 30 seconds a day or until you feel really happy with the smiles you produce, and you'll be improving your facial muscle tone. This is a

GOPex exercise an adult can benefit from as well—improving muscle tone can reduce snoring and even lessen problems with OSA.

GOPex exercises may seem like a silly series of movements that could not achieve anything dramatic. They are designed, however, to help a child or an adult hold a specific *static* position; they teach you to keep slight tension in your oral-facial system so that it develops appropriately, with good oral posture. Scientific studies have not been carried out on the results of GOPex programs. Our conclusion that they are well worth trying is based on the clinical experience of Sandra, John Mew, Simon Wong, and their colleagues.

. . .

While a young person is going through the GOPex, what other things can she be doing to encourage proper jaw development? One, of course, is to be sure to exercise her jaw muscles thoroughly. One understudied possibility is to encourage the chewing of gum.

Although many people consider chewing gum a nasty habit, and in some places it is forbidden, even illegal, in public, gum, when chewed properly with closed lips, could help children give needed exercise to their

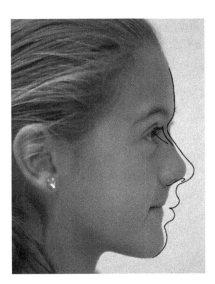 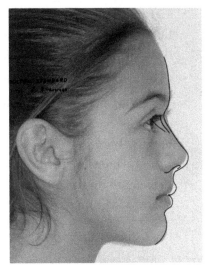

Image 53. This 11-year-old girl began practicing GOPex exercises in September 2014. Six months later her face is noticeably less concave and her lips have relaxed through the use of the exercises. It appears that her whole face has grown forward due to less tension in her lip muscles. (Courtesy of Dr. María José Muñoz.)

jaw muscles. To be successful for this purpose, gum manufacturers would have to produce products that are tougher to chew than many of today's brands, and gum that did not contain sugars, artificial sweeteners, or other materials that could contribute to tooth decay or other problems. There is a gum called Mastic developed on the small Greek island of Chios. It is a natural resin from the local Mastiha plant and demands much exercise to chew, but with some modification could be just what the dentist ordered. Developing superior therapeutic chewing gum is now on Sandra's radar. Chewing gum has proven to improve oral environment because of the well-known prophylactic effects. Chewing gum removes plaque and increases saliva, which has a protective and antibacterial effect. Chewing gum has also been linked to brain-busting cognitive performance.

Think about it: "Studies showed a professional tennis player had more bone mass and mineralization in the arm they held their racket with. Which makes logical sense." So if you exercise your jaw muscles every day, you will develop stronger, larger more powerful jaws. Chewing gum properly is something everyone can benefit from regardless of age or problem. Here's the appropriate chewing protocol:

- Chew a minimum of 30 minutes a day.
- Chew on both sides; for younger kids, supervise that they chew five times on each side. If they favor a side, consult the dentist because there may be a problem (such as a tooth decay).
- Chew with closed lips, breathing exclusively through the nose.
- When chewing, make long pauses and pay attention to how long you can keep teeth in contact as you swallow. Push your tongue solidly to the palate when swallowing. You are working on muscle memory, and you should repeat this pattern when you eat.
- Remember that this is an important exercise; it is not a habit, a hobby, or a game!

What older children and adults can do

Although this book is focused primarily on children, you should not conclude that there is nothing older children or adults can do. The primary reason for those over puberty to seek treatment is to alleviate sleep disordered breathing. Curing the problem is difficult to impossible after puberty, but much attention has been paid to the issue because so many people are afflicted, and help is available to ameliorate symptoms where complete cure is not possible. Dealing with obesity and postural issues are examples. Here we review appliances and orthodontic options as well as some thoughts on resting oral posture for getting relief for adults and older children.

Night breathing aids

In addition to CPAP machines, described in Chapter 6, a wide variety of other appliances have been claimed to help reduce snoring and some sleep disturbed breathing symptoms. These run from a vest that has a tennis ball sown in the back to keep a person from sleeping on his back, to strips that go over the nose and nasal dilators to keep the nostrils open, to oral appliances.

There has been some research on the use of Homeoblock, DNA, Oasys, and Biobloc appliances by adults; here are some of the effects that have been reported:[32]

- More prominent cheekbones
- Bigger smile
- Decreased expression lines and wrinkles
- Straighter teeth
- Better facial symmetry
- Relief of some facial pain symptoms
- Symptom relief of some mild sleep and breathing disorders
- Facial betterment (presumably more attractive!)

Just learning to keep teeth in light contact at rest can have amazing benefits in well-being as well as in esthetic improvement. This has been confirmed by Sandra and a few other adults that have focused and taught themselves to adopt this ideal oral posture at rest. Keep in mind that it is difficult, and most of the time, maxillary expansion with an appliance is needed because the tongue will not have enough room after a lifetime of keeping it away from the palate. However the sacrifice is, in Sandra's opinion, well worth it. As a patient said to her, "Knowing what I know now,

I would do whatever it takes to adopt and help others maintain a closed-mouth-with-teeth-in-light-contact-at-rest, oral posture. The muscle tone in my face has improved so much that people ask me if I had a face lift."

Airway-centric orthodontics

Some orthodontic procedures are designed to expand and enlarge the dental arches to make more room for the tongue. Depending on the severity of an adult's sleep disorder, they can be extremely successful. In some patients there is a substantial reduction of symptoms including not only better sleep but also fewer headaches and other facial pains.

Many of these procedures involve reversing the *retrusive*—pushing backward—effects of previous orthodontic treatments. For example, a 30-year-old woman was suffering from headaches and poor sleep. She had had orthodontic treatment with extraction of bicuspids (premolars—the teeth behind the canines and in front of the molars) during her early teenage years. Such extraction of healthy permanent teeth in conjunction with the fitting of braces can indeed result in beautifully straight teeth, but it has a retrusive effect on the jaws, face, and throat. As a result, this particular orthodontic strategy is now known to lead sometimes to airway problems and/or pain in the temporomandibular joint (TMJ)—the joint that connects the lower jaw to the temporal bone of the skull and, in essence, is the hinge that allows the mandible and maxilla to work against each other in chewing. Having been born and raised in an industrialized society, this patient's jaws were quite possibly already too far back even before the extractions. After she had the spaces orthodontically opened as an adult, and had four dental implants to replace the teeth that she had had removed, all her sleep-disturbed-breathing and pain symptoms disappeared.[33] Similar cases are not rare.

Considering all of the problems we have been discussing, it is hardly surprising that they frequently end up causing problems in the hinge that connects the jaws. Joints are highly adaptable, and the bones will always remodel to suit habitual rest positions. If the mouth is held open continuously, the bones of the temporomandibular joint will recontour so they join properly in that position. After that, when the mouth is closed the relationships of the teeth of upper and lower jaws will change (leading eventually to some malocclusion) and the details of where the bones join are altered. This can lead to stress, a clicking noise when the jaw operates, and

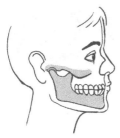

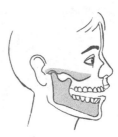

(a) When the teeth are maintained in light contact at rest, the temporomandibular joint (TMJ) will be correctly positioned and healthy. The ball on the lower jaw and the socket on the skull are properly aligned.

(b) When an individual opens his or her mouth, the ball of the joint (condyle) moves forward.

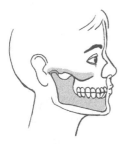

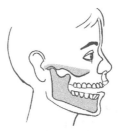

(c) After long periods in this position, the socket will reshape forward to comfortably house the ball in the open mouth position.

(d) When the mouth closes, the ball moves back and crushes the posterior part of the socket (attachment) causing pain (temporomanibular disfunction, or TMD).

Image 54. Facial pain from the jaw joint (TMJ).
When proper oral posture is maintained, with teeth in light contact, the joint maintains its proper painless configuration. Vertical facial growth, associated with this open-mouth posture, perpetuates joint stress, which can result in general headaches and difficulties talking and chewing.

pain. In its early stages this can usually be fixed with an appliance; if there is permanent damage, surgery may be required.

Correcting problems of the jaws later in life is very, very challenging and yields limited results. For that reason prevention at an early age should be everyone's goal. If the disease underlying this symptom is addressed within the first decade of life there is much that can be done to repair it and to prevent more damage. If you notice your child's teeth crowding, go to an orthodontist early, preferably someone who practices "orthotropics"

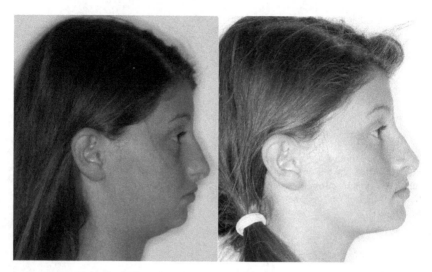

Image 55. The benefits of learning proper oral posture are observed as this girl holds her lower jaw forward.

(forwardontics). Question any advice to have teeth pulled and orthodontic techniques that move the teeth backward, and get a second opinion if possible from a health professional who will clearly answer your questions, especially on airway quality.

To make the recommendations of this chapter easily accessible, Box 4 is a short checklist of things you might consider:

When to seek professional help

If you have young children, what are the warning signs that help may be needed? Here are a few key questions to start pondering: Do your kids sleep with their mouths open or closed? Do they snore? Do they frequently have stuffy noses? Are they well rested when they wake up in the morning? The answers to such questions are rough indicators of the quality of their health that is directly linked to the shape of their face, jaw, and smile. Those answers indicate influences on the formation of the mouths with which they eat, the development of the airways that take vital oxygen from the nose or mouth to their lungs, and the shape of the faces they present to the world: their appearance.

BOX 4: CHECKLIST OF WHAT ACTIONS TO TAKE TO PROTECT CHILDREN AND ADULTS FROM THE ORAL-FACIAL HEALTH EPIDEMIC

Breastfeed for at least a year, exclusively for six months, if possible.

Avoid bottle feeding during that period, even of bottled breast milk if possible.

Do not use a pacifier until weaning is complete.

Teach children to keep their mouths closed when not eating or talking.

Lightly pinch a nursing infant's lips closed for a few seconds when she stops nursing.

Wean onto foods that require chewing, watching for choking.

Pay attention to toughness of foods, and encourage thorough chewing

Help them practice hard chewing with a tough chewing gum.

Avoid most commercial "baby foods."

Check children's sleep habits, looking for mouth breathing and signs of disturbed slumber.

Treat any signs of a stuffy nose or snoring promptly. If congestion persists, think allergies.

Encourage steps like thorough hand-washing to limit nose-cold transmission.

Check child care facilities for sanitary procedures.

Pay attention to a baby's posture, especially when it is being transported.

Limit slouching and head-forward postures over computers, phones, and the like.

Try GOPex or Buteyko training, according to perceived need.

Seek professional help promptly if you detect any of the symptoms of the oral-facial health epidemic.

In considering this issue the number one clue to how your kids are doing is this: do you see them resting with their jaw hanging, mouth open, and breathing through their mouths? If you see your child with his mouth open most of the time, that is an early indicator of potential oral-facial health problems that need attention. Watch your children when they are at rest or inactive—for example, when they are reading a book, watching TV, playing a videogame. Once attuned to the issue, you may start to notice mouth breathing more and more in your co-workers, in the driver in the car next to you in traffic, in shoppers walking around the mall. The more you look, the more you are likely to identify mouth breathing.

Another clue to good oral posture can be found in your child's smile, specifically in what we call a "gummy smile." When your child smiles, do you see a significant amount of her gums and not just teeth? That's a gummy smile, sometimes referred to as a "horsey smile." An ideal smile should display little or none of your gums. Look at your smile in the mirror; don't exaggerate, just make a natural smile. Do you see lots of your gums? If your kid has such a smile, as does the girl in Image 56, it suggests that her jaws are not developing properly, with the upper jaw growing down too far, exposing more of its gum.

Some of the clues are harder to read. Take the example of Sandra's own daughter at the age of 13. She does a good job of keeping her lips closed—

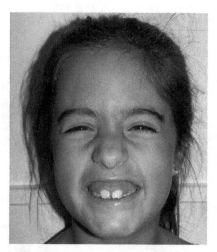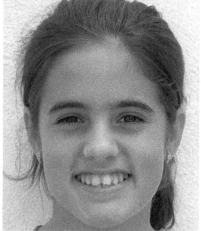

Image 56. Note the gummy smile on this seven-year old, and its disappearance after a year of orthotropic treatment. (Courtesy of Dr. María José Muñoz.)

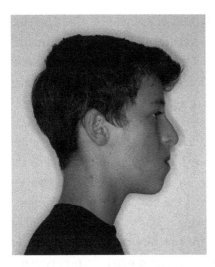 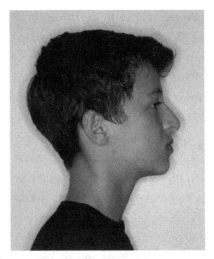

Image 57. Your child can have lips sealed and teeth apart. Many parents believe their kid has his mouth closed because the lips are sealed, but this may not be the case. Look at this boy with lips together and teeth apart on left, and on the right with teeth and lips together (proper rest oral posture). We consider the left oral posture as a mouth-open posture, even if the lips are sealed.

clue number one above—but, although she is keeping her mouth closed, her teeth are not actually touching. How can you spot this harder to detect situation? Once you know how to look for it, the surest sign of this less obvious problem is a receding upper jaw—remember, the upper jaw (maxilla), contrary to common opinion, is not fixed in place but can move gradually. Keeping your teeth apart at rest, even if your lips are sealed and breathing through your nose, can lead to snoring and sleep apnea. As we explained before, the teeth need to be in contact for both jaws to grow in unison. Some people hang the lower jaw behind sealed lips, keeping teeth out of touch and the tongue sandwiched between the back teeth of the upper and lower jaws.

What other clues should you be looking for? Watch your child swallow. Is her mouth open? Does he make facial expressions? In normal swallowing the tongue should be fully on the roof of the mouth, both in the front and at back. All the facial muscles, including those of the lips, should be relaxed during swallowing. When swallowing correctly, a wave motion of the tongue creates the required suction to move the food toward the throat and esophagus—the pathway to the stomach. The only outwardly visible movement should be in the throat. Your child's cheeks should be still. Your clue to identifying proper swallowing is directly related to how comfortable, how smooth, how seamless the swallowing motion is for your children. If you see them "squinching" (squeezing, twisting) their cheeks, making awkward movements with their tongue or tightening their lips when swallowing, it is most likely a sign of a problem with the swallowing process.

Many children do something called the reverse (or "tongue between the teeth") swallow. The reverse swallow evolved for infants who, you will recall, can both suck and breathe simultaneously. In the reverse swallow, the tongue is pushed forward, the teeth are apart, and the lips are around the tongue, which is positioned low. Reverse swallowing is a response to early weaning and weaning to semiliquid foods that don't require chewing. It ordinarily starts to fade at around six months. Some children never learn to swallow like adults. You can tell if a child is still reverse swallowing: if her facial muscles are in action while she swallows, or if she starts her swallows with her teeth apart, she is likely reverse swallowing.

What problem does poor swallowing cause? A correct swallow stimulates the palate (maxilla, roof of the mouth) to grow upward and outward, widening the dental arch and making more room for the teeth. A reverse swallow has the opposite effect, not stimulating the dental arch to expand and thus in effect crowding the teeth,[34] enlarging the cheek muscles through

Image 58. If you see your kid's face muscles in action when she's swallowing, that is yet one more sign that things are not right. Notice the pursing of the lips as this child swallows.

exercise, making the face bulge, and removing the dimples people find attractive in Hollywood models.

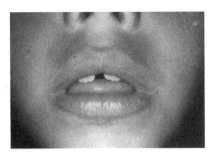

There are many other clues to potential oral-facial growth problems. Does your child show a lot of gum when he smiles and have droopy eyelids, or does she have a strongly arched upper lip (known as a "Cupid's bow")? Traditional dentists and orthodontists may look at

Image 59. If a parent sees his or her child's tongue when he's swallowing, that is an alarm bell.

just the smile, the bite, and how straight each tooth is in relationship to the others, but there is much more to oral-facial health. For the range of signs of potential problems, see the checklist in Box 5.

BOX 5: CHECKLIST – SIGNS OF PROBLEMS IN ORAL-FACIAL HEALTH

While your child is sitting (watching TV or in the car) does he or she:
- Put things in his mouth (toys, sleeves, pencils, fingers, etc.)?
- Suck her lips?
- Have an open mouth, even if only a little?
- Have his tongue between his teeth?
- Rest her face on her hand?
- Breathe through his mouth?
- Make a noise while breathing?
- Have a hard time sitting still?

While your child speaks does he or she:
- Talk too fast?
- Talk too slowly?
- Stop to breathe through the mouth?
- Lisp?
- Do the lips only rarely come in contact during speech? Ideally, lips should touch between each word.

While your child eats does he or she:
- Stop to breathe through the mouth between mouthfuls?
- Stick her tongue out when swallowing?
- Stick his tongue out when drinking?
- Drink a lot of liquids with her food?
- Make a lot of noise when chewing?
- Take a breath when drinking?

- Tighten his lips when swallowing?
- Wrinkle her chin when swallowing?
- Tilt his head when swallowing?

While your child is sleeping does he or she:
- Sleep with her mouth open?
- Snore?
- Wet the bed?
- Toss and turn?
- Stretch his head backwards?
- Wake up frequently?
- Have nightmares?
- Grind her teeth?
- Have trouble waking up?
- Have dark circles under his eyes?
- Wake up drooling or with dry saliva on her face?

Why these symptoms occur is not always obvious, nor are they all necessarily traceable in any individual case to issues of oral-facial health. But many are, even when the connection seems strange. Why, after all, should bed-wetting be connected to the jaw? But the evidence that it is is that if adequate room is made for the tongue and the nasal airway improved, bed-wetting ceases.[i] The story with nightmares is similar.[ii]

i D. J. Timms. 1990. Rapid maxillary expansion in the treatment of nocturnal enuresis. *The Angle Orthodontist* 60: 229–233.
ii P. Jaoude, L. N. Vermont, J. Porhomayon, and A. A. El-Solh. 2015. Sleep-disordered breathing in patients with post-traumatic stress disorder. *Annals of the American Thoracic Society* 12: 259–268; B. Krakow, C. Lowry, A. Germain, L. Gaddy, M. Hollifield, M. Koss, D, Tandberg, L. Johnston, and D. Melendrez. 2000. A retrospective study on improvements in nightmares and post-traumatic stress disorder following treatment for co-morbid sleep-disordered breathing. *Journal of Psychosomatic Research* 49: 291–298.

CHAPTER 8

ORTHODONTISTS, DENTAL ORTHOPEDISTS, ORTHOTROPISTS, AND FORWARDONTISTS

If you decide your children (or you) need help with a problem of oral-facial health, where should you turn? Which health care professionals are most likely to produce desired results for you, and what can you expect of them? How easy will it be to get the help you and your children need?

Naturally, you will first think of consulting an orthodontist. The long-established practice of orthodontics has increasingly become a routine part of the normal journey of childhood, at least in the industrialized world. Remember that it's estimated that well over half of American kids are now using braces sometime in their development.[1] That is partly because the cost of orthodontics has dropped dramatically, partly because of a rise in self-consciousness about not having picture-perfect straight teeth, but almost certainly also because children increasingly are not developing straight teeth naturally.

There is little doubt the severity of orthodontic malocclusions is increasing.[2] The number of children having teeth out and the number of teenagers having surgical jaw corrections is on the rise, despite the improvements in the wire and bracket technology of braces. Further, there's understandably increased concern with breathing function, which, as we have seen, is tightly tied to jaw size and structure and thus to tooth straightness.

Here is a rundown of the essentials of the profession of orthodontics and two of its branches, dental orthopedics and forwardontics and the treatments each specializes in, and how each may or may not play a positive role in the oral-facial epidemic.

Orthodontists

Orthodontists were originally trained as dentists, so, not surprisingly, they are focused on correcting the arrangement of teeth. Frequently their main goal is a smile showing even rows of teeth, with no overlapping teeth on the same jaw, no teeth with unusual orientation, and no upper teeth so far in front of lower teeth that the latter are nearly absent from the smile. A major way orthodontists achieve their results is by deploying various devices, such as braces, that tug on the teeth, gradually moving them through living bone into more desirable positions.

In their training, orthodontists are taught that facial features are mostly inherited; to see this all they need to do, they're told, is compare the fea-

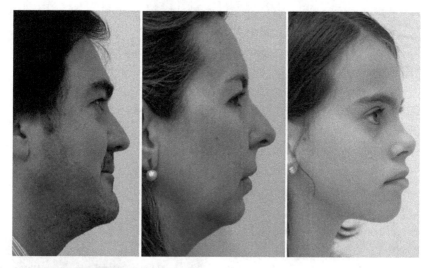

Image 60. Father and mother had traditional orthodontics. They chose forwardontics for their daughter. Notice the different angulation of the mother's and daughter's front teeth. (Courtesy of Dr. María José Muñoz.)

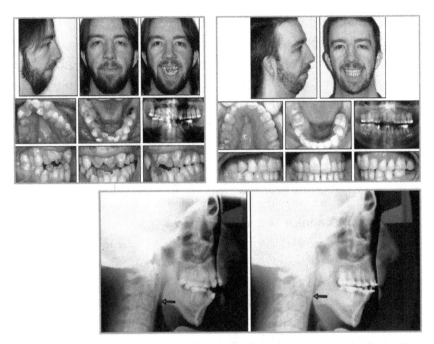

Image 61. Case (before and after) used to illustrate outstanding orthodontic results in a peer-reviewed journal. Note the near total constriction of the airway after treatment (bottom right). (*Journal of Clinical Orthodontics*)

tures of relatives within and across generations. Many people are under the misimpressions that genes are destiny, that there is no way of altering that destiny, and that all one can do is make cosmetic changes, much as a hairpiece is used to carpet the head of a person with male-pattern baldness. We can't emphasize too much that genes cannot even be thought of without consideration of the environments in which they operate, and that in the absence of appropriate cellular and external environments serious defects in the final living product occur. We can't repeat too often that too small jaws and crooked teeth trace to inappropriate environments in which genes are expressed.

It is thus tragic that in general orthodontists think that children's dental crowding—jaw and tooth size mismatch—is inherited from their parents. Many orthodontists typically think working with the jaw in early years is not only a waste of time but likely a fraudulent practice.[3] The crookedness, in their view, will develop whether anything is done or not, so therapy is limited to cosmetic management. This DNA-centric misapprehension is one reason why standard orthodontic procedures focus on correcting the symptoms (misaligned teeth and jaws) rather than addressing their frequent nongenetic underlying causes. These include, as you are now hopefully aware, lack of adequate chewing, bad oral posture, high-allergen environments, and the like. These causes, we reemphasize, should be addressed as early as possible. To not treat the issue early would be the equivalent of detecting high blood sugar in a young child, and, as pediatric dentist Kevin Boyd points out,[4] is like saying let's not treat the child until he's fully diabetic. This faulty delayed treatment is connected to a general society-wide fascination with DNA and the notion that it alone is the most powerful actor in human biology.[5]

The overwhelming majority of orthodontists practice teenage fixed-braces therapy with surgical operations on the jaw after puberty for patients with severe problems. This style of practice is all-but-universally accepted as the normal standard of care and reveals the orthodontic profession's underlying belief in the myth of an overwhelmingly genetic origin of malocclusion. Relatively late treatment makes sense to them, as the changes that accompany growth have slowed considerably by the teenage years and thus growth, the predominant source of tooth movement, no longer needs to be considered. Crooked teeth in older kids are less of a moving target for correction, and treatment outcomes are more readily predicted.

Appropriate long-term research, however, shows that the standard orthodontic correction of misalignment is typically transient.[6] For instance, one detailed study of records of more than 800 cases at the University of Washington by Dr. Robert M. Little, specialist in long-term effects of orthodontic treatment, concluded that after orthodontic treatment room for the tongue shrinks and malocclusion returns, but the degree of relapse varies from individual to individual.[7] It is rare to find children who have had orthodontic corrections and maintain the straightness achieved into adulthood.

In John Mew's identical twin study (which we describe in the following discussion), observations of jaw growth in family members in different environments, long-term research on the stability of postretained orthodontic results (that is, how well the teeth stay in place after treatment), and, especially, basic evolutionary theory make it crystal clear that claims of a dominant role of genetics in malocclusion can be ignored in almost all cases. Hard as it is for the profession to admit, there is little in the way of hard science behind orthodontic practice.[8] That is not for lack of interest, of course: the very nature of working on people, especially children, entails ethical restrictions that preclude many scientific studies on human beings.

Dental Orthopedists

A small group of orthodontists, the dental orthopedists, along with some pediatric dentists are oriented to treat younger children starting around the age of four or five. These clinicians believe certain environmental factors can cause imbalances in development of the teeth, jaws, and face. Practices such as lisping in speech, mouth breathing, and such oral habits as lip and thumb sucking, in their view, lead to crooked teeth and related problems that need to be dealt with at an early age.

Their techniques are based on a recognition of the origins of a problem in the jaws, and they attempt to place the jaws in the ideal position using bulky appliances that do not allow the patient to continue to do the movements that caused the problem (like reverse swallowing). The appliance does the work for the patient. Sometimes exercises are prescribed to help change the muscular habits; these exercises are difficult and boring, so orthopedics relies more on the appliances. Muscles end up fighting the appliances, and the result is usually unsuccessful.

Choosing the orthopedic route is attractive to some parents because orthopedists start treatment early and promise to fix some of the developing problems that the parents notice in their kids. Orthopedic practitioners frequently promise as well that permanent teeth will not need to be removed because the orthopedist will expand the jaws to make room before all the permanent teeth erupt.

Sandra practiced dental orthopedics for over 20 years, and she has observed the results on patients. Twin Block and Herbst, two of the most popular orthopedic appliances, set up a slack-jawed child to correct the underdeveloped lower jaw. They hold the lower jaw forward with a device anchored to the upper jaw (maxilla). The problem, which very few people recognize, is that the upper jaw is not a fixed block of concrete to which one can anchor the wayward lower jaw without consequences. These orthopedic appliances, in our view, can in effect use the lower jaw to drag the maxilla down and back to meet it, often constraining the airway in the process.

We believe another questionable orthopedic technique is the use of cervical headgear (headgear that anchors on the back of the neck, not the headgear that pulls forward) which either just holds the upper jaw in place, not permitting it to grow as it should, or worse, pulls it back,[9] thus constricting the airway.[10] In addition to making children suffer discomfort and embarrassment, we think cervical headgear only makes their situation worse.

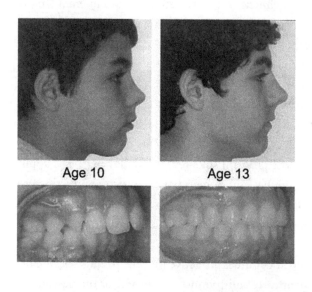

Age 10   Age 13

Image 62. Note the improvement in the teeth alignment, although the face has elongated and become somewhat concave at 13. This shows the focus on teeth alignment rather than facial structure. From *The Journal of Clinical Orthodontics*.

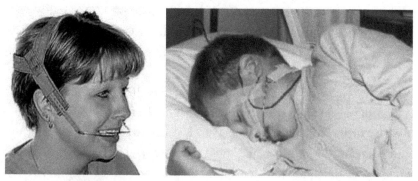

Image 63. Cervical headgear used in traditional orthodontics pulls the jaw back which can result in constriction of the airway and OSA. Traditional Headgear. Research showed that the use of retractive headgear increased sleep apnea. (*Journal of Pediatric Dentistry*, 1999)

Even with early orthopedic treatment, teeth alignment often also needs later intervention, in the majority of cases with braces in the teenage years, after the permanent teeth erupt. Peer-reviewed research studies tend to reveal little improvement in outcomes with such early intervention functional therapy when compared to just late intervention with fixed braces alone, as a fine review by Lysle Johnston demonstrates.[11]

Forwardontists

The smallest subdiscipline of orthodontics, forwardontics represents a break from the traditions of both standard orthodontic and dental orthopedic practices. If begun early enough, orthotropic treatment can even result in complete cures, as the clinical work of John Mew, Simon Wong, and others has shown.[12] Forwardontics is the new kid on the block in oral health, however, and its advantages are not widely recognized, nor its therapeutic techniques yet widely practiced, among health professionals.

Forwardontics (also frequently called Biobloc or facial orthotropics), in contrast to the remainder of orthodontics, focuses on the face and the causes of malocclusion. It is most often practiced by some general dentists who, having seen the disappointing results of standard orthodontics (cases of facial damage, airway restricting, pain, lack of long-term alignment stability, and so on), have embraced this therapy as a viable alternative.

(b)

(c)

(a)

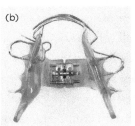
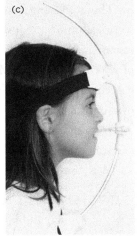

Image 64. (a and b) Mew's Stage 4 Biobloc: the legs train patients to keep their teeth in contact "voluntarily" because they are uncomfortable when teeth are separated. (c) BOW: forward traction without facial anchorage. These are forwardontic (orthotropic) appliances that expand, move the teeth and both jaws up and forward, and train the patient to keep his or her mouth closed.

Forwardontics, like standard orthodontics and dental orthopedics, is concerned with how the teeth are arranged, but it works to correct problems with the teeth in conjunction with those of jaw size and facial structure. It is especially concerned with avoiding the serious problems that can develop in the airway. Although forwardontics also uses devices, it works to avoid malocclusion by training people to counter the malign effects associated with those aspects of our industrialized existence already discussed, by adopting a healthy oral posture, chewing more thoroughly, reducing nasal congestion, and so on. It works to return the growth pattern of the face to the evolutionary path of our ancestral past.

The forwardontic approach can be largely traced to the British dentist John Mew, who, you'll recall, called it orthotropics. In the 1970s, after much experience treating malocclusion and observing how children grow, Mew came up with what he termed the "tropic premise,"[13] the word *tropic* referring to growth in response to a stimulus. Many problems of malocclusion, he argued, could be remedied if young children were encouraged to practice what he believed was proper resting oral posture as the stimulus, the now-familiar with "tongue resting on the palate, the lips sealed, and the teeth in light contact for between four and eight hours a day." Based on this premise, Mew developed a set of orthotropic appliances that could aid in restoring such proper resting oral posture. The appliances are uncomfortable when the mouth is hanging open, so they "persuade" patients to keep their mouth closed by encouraging the use of their own jaw muscles (an improvement on George Catlin's suggested remedy of "strapping up the jaw"!).

This active engagement of the orthotropic patient in the treatment is a critical point, because all other orthodontic and orthopedic appliances essentially do the work for the patient and don't engage jaw muscles, use of which is central to Mew's orthotropic program. Mew's ideas were fundamentally simple and based on many years of working with patients, his own critical studies of identical twins, and of course on the research other scientists and practitioners had published on the importance of determined chewing in development of the jaw and face.

Mew's basic theories can be easily summarized:

1. In practically every person in modern society both the upper jaw (maxilla) and lower jaw (mandible) are well behind their ideal for-

ward locations for airway development even if the upper jaw appears to protrude.[14]

2. Such underdevelopment of the jaws, as we have seen, is typically a consequence of the person's eating and breathing history causing an improper interaction between the tongue, the roof of the mouth, and other muscles and bones in the development of the face and jaws.

3. The position of the teeth is not static; teeth keep moving slowly throughout life.[15] Bone is always being dissolved and reformed—reorganized or "remodeled"—contrary to the common impression that it is static and permanent.[16] Thus its shape and the direction that teeth can be gradually moved through it can be changed, This flexibility, the point that bone is not cast like concrete, is the basic fact behind both orthodontics and forwardontics.

Mew's aim has been to cure the problem rather than, as standard orthodontics often does, simply putting it in remission. Malocclusion always returns, at least to a degree, after standard orthodontic treatment ends, if a "retainer"—a device that holds the teeth in their new position—is not worn, according to one of the few studies of the subject.[17] The standard of care in orthodontics is thus to instruct the person to wear the retainer for-

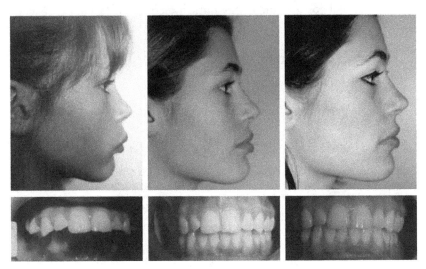

Image 65. Orthotropics protects or even enlarges the airway by directing the growth of upper and lower jaws forward as a unit. (Courtesy of John Mew.)

ever. This is in line with the general trend of modern medicine to focus on the maintenance of chronic diseases, rather than dealing with their causes. Judging whether problems of oral-facial health have been permanently fixed requires viewing photographs of patients taken before treatment and at least five years after retainers have been removed,[18] and that is not commonly done. Teeth do not wander after orthotropic treatment because properly related jaws, teeth, tongues, and lips serve, in essence, as retainers.

We believe Mew's pioneering work deserves close attention by anyone in charge of patient well-being. It is clear that the direction of facial growth, so important in the development of malocclusion and later problems such as sleep apnea, is not inherited because it varies a lot even in identical twins and diverges in close relatives exposed to different environments. All aspects of human development, of course, result from the interactions among genes and interactions of genes with environments. Just as one identical twin can, through exercise, become a fine athlete while her genetically identical sibling may excel as a sedentary artist, so one twin eating a diet of soft food may turn out less healthy and less likely to be considered attractive than his "identical" sibling who has done a lot of chewing.

An experimental study Mew carried out with identical twins ranging in age from 8 to 19 years provided powerful evidence that facial growth direction varies considerably with environmental conditions, including patterns of chewing, oral posture, and orthodontic intervention.[19] Mew approached

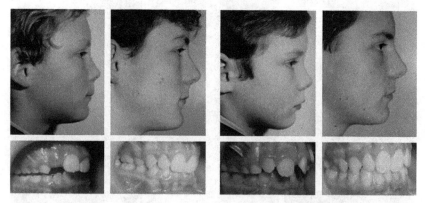

Image 66. One of the sets of identical twins who participated in Mew's study. Quinton (R) was treated by Mew, whereas Ben (L) was treated by a local orthodontist. The results of the two types of treatment over a span of 12 years are markedly different. Look at the fullness of Quinton's profile in comparison to Ben's dished in longer face.

the parents of six sets of identical twins and told them, "I will treat one of your children at no charge if you agree to take the other to a local orthodontist, and you agree to let me document what happens by taking pictures of both children. Please tell the orthodontist this is a contest to see who can produce the best looking face and that he/she also gets to choose which twin he wants to treat."[20]

Ten years later a panel of experts evaluated the sets of twins. The results were conclusive, even though the sample size was necessarily small. Overwhelmingly, the judges rated the traditionally treated twins less attractive after treatment than before, and their twins, who received Mew's treatment, which included training in oral posture as well as orthotropic intervention that employed appliances that were not fixed, more attractive after treatment than before. Moreover, all the twins Mew treated had teeth that remained straight for at least a decade, which is not surprising because, as we said, proper oral posture itself functions, in essence, as a retainer. This experiment showed, above all, that genes are not destiny—individuals with identical genes became different in different treatment environments.

Forwardontics (orthotropics) as developed by John Mew was designed to prevent malocclusion by guiding growth from an early age, restoring the patterns of chewing pressures and resting posture impacts on muscles and bone typical of traditional human societies. The idea was to keep your mouth closed and to build core strength within your mouth, so that the

Image 67. At maturity, Ben on the left has a thinner lips and a longer face than Quinton on the right. The orthotropic treatment was judged by experts and lay people to be superior.

upper and lower jaws are matched together and growth is symmetrical, balanced, and forward, resulting in a bigger, wider jaw.

Forwardontics develops a position for your mouth to hold (good oral posture) to counter the forces of gravity that tend to pull the upper jaw down, just as developing overall good posture allows us to oppose the forces of gravity bending us over.[21] In ideal oral development, the lower jaw supports the upper jaw against the pull of gravity through the work of muscles, including the tongue, toned by appropriate weaning, chewing, breathing, and so forth. Remember, despite casual impressions, the upper jaw is not fixed and would tend to slide back and down without gradual muscular pressure.

Environmental pressures in modern lifestyles are not conducive to generating the needed muscle tone, however. Consequently, Mew developed the Biobloc, the device previously mentioned for children to wear in their mouths to encourage them to maintain correct oral posture. It not only requires patients to actively keep their mouth closed but also guides their jaws in growing to the ideal conformation.[22] When used under expert supervision and on cooperative patients, the device has produced some impressive results,[23] as shown in Image 68.

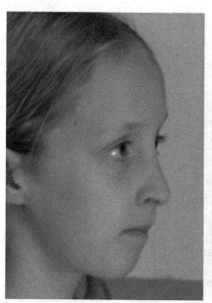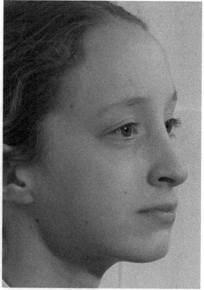

Image 68. Young girl who chose orthotropic treatment instead of surgery to improve her receding chin. The success of the treatment is clearly visible in the "before" and "after" images. (Courtesy of Mike Mew.)

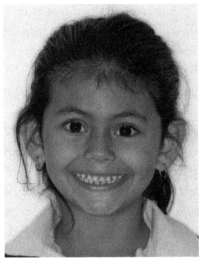

Image 69. This 4-year-old girl had no spaces between her baby teeth. She lifted up her chin to be able to breath better. She went through 6 months of expansion and postural work, and the results show how important is to treat children early. (Courtesy of Dr. Simon Wong.)

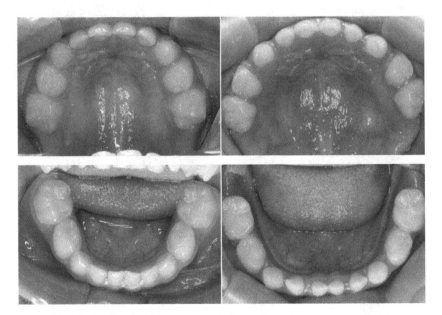

Image 70. Baby teeth should have spaces between them to accommodate the permanent teeth without crowding later. See the changes with 6 months of expansion and postural work. Look at how happy the tongue is in the right side!

Another important tool of forwardontics is the GOPex set of exercises (described in Chapter 7) to help the patient develop a proper posture when the mouth is at rest. Exercises can be tedious (for both parent and child), but typically no surgery or extractions are required, and the results in improvement of health and facial appearance are often dramatic. However, forwardontics depends primarily not on the appliances themselves but on serious interest in appropriate life-changing behavior.

How does GOPex play a central role in forwardontics? Dentistry in general is largely a hardware-based therapy: a crown, a filling, a tooth extraction, braces, brackets, headgear, and so forth. When dentists provide some hardware and ask patients to complement with exercises, many patients get lost in the hardware and expect that the change will result from use of the appliance or the brace. Patients often understand the concepts that muscle development and resting posture are important but do not consider them as significant as the hardware. Motivation behind doing the

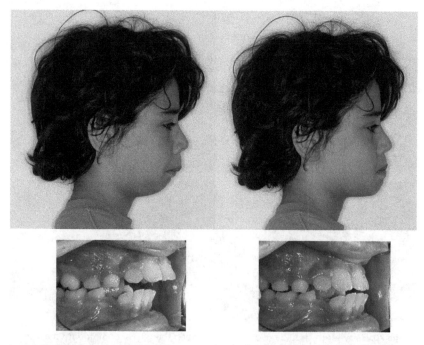

Image 71. Photographs taken on the same day, before and after a patient is instructed on how to hold his lower jaw forward. With long-term proper postural exercises, skeletal improvement can become permanent. (Courtesy of Dr. Simon Wong.)

simple GOPex exercises may disappear when normal life catches up. As Dr. Simon Wong emphasizes to the parents of his patients, "When you pay me for my services, I am training you to retrain your child to develop their body to their best potential. The exercise program is the fundamental component of what you are paying for; I am throwing in the hardware and the retainers for free."[24] This is why forwardontic treatment depends heavily on patient compliance and not on the appliances themlseves. In contrast. the orthodontists' braces don't get bored or busy.

Adult Forwardontics

There is a pioneering field of adult forwardontics in which adults are fitted with the same type of appliances that kids are given with some good initial results being reported.[25] Sandra has used two standard forwardontic appliances successfully with adult patients with mild symptoms of sleep-disordered breathing. Both look like orthodontic retainers and are worn only at night. The Homeoblock (similar to the much-used DNA appliance)[26] and the oral-nasal airway system (OASYS) direct tooth movement to widen the dental arch and advances the lower jaw, making more room for the tongue and tending to open the nasal passages, improving the capacity of the airway for snoring and mild sleep apnea victims. Only time will tell if long-term good results, especially more open airways, can be expected from treatment of adults.

Forwardontics and orthopedics: The critical differences

Patients often confuse forwardontics and dental orthopedics, in part because both emphasize early treatment, but the differences between the two techniques are significant, as are their treatment results. For example, expanding devices for the upper jaw in orthopedics and forwardontics look similar, but the effects are quite different. Forwardontic expansion is typically more effective and more stable over the long term because of the growth-guiding and behavior modification component of forwardontics.

Compared to orthotropics and forwardontics, which focus on growth direction through training, dental orthopedics focuses mainly on changing the lower jaw—recall that orthopedic appliances can drag the upper jaw down while trying to move the lower jaw forward. Although only insufficiently expanding the upper jaw and not moving it forward, orthopedics anchors devices from the upper jaw and actually may push the it back, potentially causing serious side effects. Only in cases where the teeth of the lower jaw protrude more than those of the upper jaw will dental orthopedists use re-verse-pull headgear, anchoring mostly from the lower jaw. Many of these cases of lower jaw protrusion end up needing surgery because dental ortho-pedic practice does not usually treat for enough time or succeed in changing the habits that produced the problem in the first place.

It is no coincidence that Dr. Lysle Johnston, the conscience of the orthodontic community, entitled his key review paper, "Growing jaws for fun and profit,"[27] showing that dental orthopedic appliances typically do not work as intended.[28] They can help move teeth around temporarily, which can help with appearance, but have little long-term beneficial effect. Growth takes many years; if permanent improvement is to be obtained, a fundamental change in the patient's oral posture is required, a critical point dental orthopedics overlooks. The behavior that caused the problem must be extinguished, or the problem will return; continued treatment along the same lines would become an example of a definition of insanity usually at-tributed to Albert Einstein: doing the same thing over and over again and expecting different results.

In curing malocclusion, the upper jaw almost always needs to be en-couraged to move *forward*, and that is the foundation of forwardontics. As we've indicated, the upper jaw is considered by too many orthodontists to be fixed, which it certainly is not. The upper jaw's mobility is demon-strated, among other ways, by the very "success" of dental orthopedics in moving it back. Forwardontics focuses on starting treatment in the first de-cade of life and on widening the upper jaw while moving it up (shrinking the distance between the upper lip and nose) and forward. If the patient maintains proper oral posture, that change will be permanent.

Orthodontics and health of the airway

Orthodontics and dental orthopedics may improve the positioning of your child's teeth, but they may have negative impacts on health. For nearly a century, many aspects of standard dentistry, including fitting dentures, mak-ing crowns, and doing extractions, have had the side effect of moving structures backward and reducing space in the mouth, making it more likely that the tongue will move backward in the mouth and reduce the path air must follow to the lungs.

Sandra has seen this firsthand many times, having treated roughly 2,000 people over several decades using standard orthodontic techniques. She grew increasingly uncomfortable because she was seeing remissions, not cures. Her tipping point came when her son Ilan began to show signs

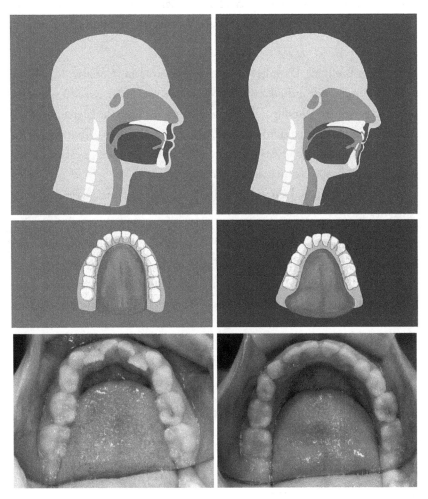

Image 72. In the top two rows, you can see how extraction of teeth reduces the available space for the tongue. In a smaller arch the tongue will drape back into the throat causing snoring or sleep apnea. Note in a different patient below, there is more room for the tongue after orthotropic expansion. (Courtesy of Dr. María José Muñoz.)

of problems with his airway—snoring and mouth-open posture. That is when she discovered and then became involved in the practice of orthotropics and eventually renamed the program forwardontics. She came to the conclusion that forwardontics was the only treatment that changes facial development in a way that increases the size of the airway and in turn prevents snoring and sleep apnea.

It isn't as if orthodontists have been unaware of airway problems, though, and of the frequently temporary results of their efforts. Many orthodontists recognize the scarcity of solid scientific evidence for their procedures.[29] In a recent review of a massive literature,[30] orthodontist Ki Beom Kim stated:

SBD (Sleep Breathing Disorder) is not a simple problem. However, it can be a great opportunity for orthodontists to collaborate with other medical specialties to improve a patient's health and treatment outcome. Since the beginning of our specialty, there has been a continual progression of improved understanding of the link between function and facial growth and development. Future research presented in our scientific journals in the next century may shed a light for us to better identify the problem and aid our specialty in developing more effective evidence based treatments.

Orthodontists of course are well intentioned, as Kim pointed out: "Orthodontists today still agree with the following statement written by M'Kenzie:[31] 'Thus in your orthodontic work you are engaged in a great labor for the prevention of disease not only in childhood but also later in

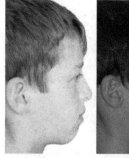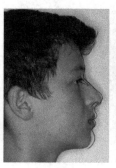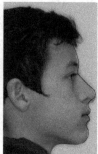

Image 73. Forwardontic treatment from age 10 to 16 years.

life, and your efforts, when successful, make for the prolongation of life.'"[32] But being well intentioned does not necessarily produce good results.

Recently a visionary group of dentists and physicians have increased the focus on the airway. They have discovered that such things as pain in the temporomandibular joint that connects the lower jaw to the skull, snoring, and some headaches and other chronic complaints disappear if attention is paid to enlarging the airway. They have formed a group called American Academy of Physiological Medicine and Dentistry (AAPMD).[33]

Donald Enlow, an authority in facial growth and development and author of *Essentials of Facial Growth,*[34] put the key concept concisely: "The airway is the cornerstone of facial growth." The AAPMD, of which Sandra is a member, represents a new trend in which airway-centric practitioners agree never to choose a treatment plan that shrinks space for the tongue and thus for the airway, even if it might be harder and take longer. They don't start with a focus on the position of the teeth, but rather with a focus on breathing, and construct their treatment with that as a priority. Forwardontics is an airway-centric therapy, but strives not only to protect the

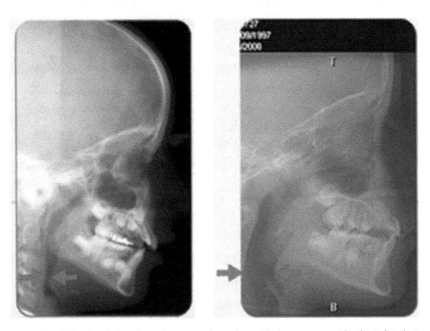

Image 74. Orthotropic treatment can produce dramatic improvement in the tube that takes air to your lungs. (Courtesy of Dr. William Hang.)

BOX 6: DIFFERENCES BETWEEN ORTHODONTICS AND FORWARDONTICS

Orthodontics/Orthopedics	Forwardontics = Orthotropics
Moves teeth	*Changes patterns* of facial growth
Start evaluation at 6 years old	Start evaluation postnatally
Phase I around 9 years old	Active treatment at 7 years old
Mostly fixed appliances	Mostly removable appliances
Some removable appliances	Limited fixed appliances
Retrusive mechanics (pushes teeth back)	Protractive mechanics (moves teeth forward)
Focuses on the mandible (bottom jaw)	Focuses on the maxilla (upper jaw)
Pushes upper jaw back (cervical headgear)	Grows upper jaw forward (reverse headgear)
Sometimes requires extractions	No extractions of permanent teeth
Most often wisdom teeth need removal	Creates normal space for wisdom teeth
Tendency for relapse of crowding	Sometimes extra space is left over
2 to 3 years active with lifelong retainer wear	2 years active;semiactive until growth is complete, but no need for retainers
Cost: about US$6,000 (2015)	Cost: about US$15,000 (2015)
Ignores the airway	Enlarges the airway
Ignores the proportions of the face	Changes the face to appear more full
Mechanical in nature	Postural in nature; Retrains muscles
	Trains muscles to rest the jaws properly
Orthodontist does most of the work	Highly dependent on patient cooperation

airway but also to improve it by regaining the ideal rest oral posture, thus ensuring the evolutionarily preprogrammed forward growth of the face.

Change and the Orthodontic Profession

Sadly, Mew's views are not widely accepted and acted on by those dedicated to oral health. Why haven't more orthodontists focused their attention on issues of oral posture, diet, and airway considerations? First of all, American dentistry tends to be highly specialized—or "siloed," in current jargon. Some practitioners "drill, fill, and bill," whereas others move teeth around. The discipline as a whole does not emphasize more broad-based treatment that considers all aspects of the patient, including his or her behavior.

Additionally, professionals in every field, not just health, tend to be very conservative, even sometimes in the face of massive contrary evidence. Even the best of practitioners have sometimes been slow to adopt lifesaving procedures. A classic case was that of Ignaz Semmelweis and childbed fever. Semmelweis was a Hungarian physician who showed in the 1840s that the dangerous fever, which killed about 10 percent (in some cases, up to 30 percent) of women in maternity wards, could be nearly eliminated by careful hand washing by physicians. Doctors largely refused to believe his empirical evidence because it conflicted with then-current theories of contagion, which did not yet include Louis Pasteur's germ theory. They recognized that diseases could be transferred from person to person, as Edward Jenner had shown by spreading cowpox from individual to individual to "vaccinate" them against smallpox. But they had no idea what the causative agent was, and many still believed that small animals like flies could be spontaneously generated from nonliving things like garbage. Semmelweis himself came closer than most to the truth, thinking it involved transfer of minute particles of corpses when doctors did autopsies and then did not wash their hands before treating patients. Semmelweis also threatened the doctors' self-image by implying physicians themselves could be a source of harm. So women continued to die for several more decades, until Pasteur and others demolished the old contagion theories once and for all. What was once mysterious is now recognized as the result of bacterial infection of the female reproductive tract.

Another reason for orthodontic reluctance is that conventional evidence on how the jaws and face evolve and develop is difficult to obtain, though the same might be said of the orthodontic literature more generally. It is sadly impossible to compare hundreds of sets of identical twins like Mew's, with one of each pair bottle fed, weaned onto commercial baby food, and given a standard industrial diet and the other nursed for two years, lips pinched shut after every feed, weaned gradually onto chewy food, and then fed a diet that minimized highly processed foods. But obviously less well-controlled studies must be substituted—comparing people who have moved from traditional to industrial societies with those who stayed home, examining skulls from different preindustrial societies and comparing them with those from modern developed countries, and the like. In addition, there is substantial clinical evidence in support of an forwardontic approach simply in the results forwardontists often get in their practices. In areas as diverse as cosmology, ecology, paleontology, climatology, and human behavior, such nonexperimental approaches have provided rich insights. Like most physicians, though, many dentists have little or no training in evolution or in how research is properly done. For example, in the twin study Mew was told by colleagues his sample was "too small," when, in fact, it was not necessarily too small at all; comparison of just one pair of identical twins can sometimes yield interesting information.

A related factor is a lack of basic understanding embedded in the techniques and goals of orthodontics. As an example, we have already noted that extractions can lead to reduced airways. Some papers have claimed that extractions have no effect, but the research designs used to arrive at that conclusion have been shown to be flawed.[35] They examine the airway (by x-ray) prior to extraction, wait a couple of years, and reexamine it (by x-ray) and find it hasn't changed. But, to get a true gauge of the potential effects, one needs to examine x-rays taken 20 years after the tooth removal. Extraction usually leads to a gradual shrinkage in the volume of the mouth,[36] reducing space for the tongue and making it more likely to fall back when someone is sleeping on her back. Despite continuing defensive attempts to exonerate orthodontics from responsibility for the onset of OSA,[37] we now have multiple studies with solid evidence that link the loss of individual teeth or the smaller size being unequivocally linked to OSA.[38]

It is difficult for orthodontists, as it is for all of us inhabitants of a fast-paced society, to focus on long-term effects. And like so many people in our society, often medical professionals have not been educated to appreciate the environment–health big picture, which is largely focused on helping "cure" the diseases of individuals rather on removing the underlying factors generating ill health. Unfortunately, in standard orthodontic schooling, little attention is paid to airway issues or other related oral-facial issues.

Finally, changing to a new approach in orthodontics, as in many other professional fields, often requires considerable time and effort and may not be in the short-term financial interests of well-established practitioners.[39] Asking people to change their well-established ways, especially when the change may cost them money, is generally met with great resistance. The situation with CPAP machines and surgery for impacted wisdom teeth throws light on how financial issues can sometimes influence health care decisions. If one has invented an excellent CPAP machine and has a financial interest in its manufacture, it would take a superhuman effort not to be biased toward prescribing its use or to encourage long-term solutions that would cut into its market. And oral surgeons clearly have an incentive to ignore the news that their cash cow of extracting wisdom teeth prophylactically may actually contribute to oral-facial problems.[40]

Selecting the right health professional to deal with suspected oral-facial problems is no trivial challenge. There's no easy rule, just a series of issues that should be discussed with a candidate before you make a choice. By now you should know most of the questions, about how cosmetic treatment undertaken will influence the airway, about the practitioner's views on extractions, and so on. If you are lucky enough to find a forwardontist when your kids are still young, the biggest question may be one you ask yourself—am I ready to make an enormous investment in time and money if my child needs help, and is he or she willing to undertake a long program dedicated to curing the problem? Time in selection will be well spent, especially because you may be committing your child to a decade of treatment (2 years active, followed by perhaps 8 years semiactive) and you to both supervision and perhaps tens of thousands of dollars of expenses (depending on insurance, number of children, geographic area, and your selection of professional).

CHAPTER 9

CHANGING CULTURE,
IMPROVING HEALTH

The predominant cause of malocclusion, in the forwardontic view (and ours), appears to be an open mouth too much of the time in childhood and failure to maintain proper oral posture for extended periods, especially when asleep. This, along with habitual mouth breathing, possibly less breastfeeding[1] and new patterns of weaning,[2] modern diets and eating implements that impede development of proper muscle tone by chewing hard,[3] and living mostly indoors are major factors in causing distorted facial development. These problems are part of a series of large-scale and interrelated difficulties ranging from overpopulation and wasteful consumption to climate change facing our children, our grandchildren, and ourselves. More specifically, they are part of health problems worldwide in which the predominant focus is not on prevention but on management, so much so that in the United States and much of the rest of the world it would be more honest to call the "health care" system a "health repair" system (at best).[4] It's the difference between fostering wellness and the more typical policy by default of managing disease as an afterthought, which, in the United States[5] and elsewhere,[6] creates a gigantic economic burden. That huge cost could be greatly reduced by giving primacy to prevention.

In matters of health, the principle of "an ounce of prevention is worth a pound of cure" once again shows its validity, as it does in most of the other environmental problems of what has become known as the "human predicament"[7] society faces today. Climate change would have been infinitely cheaper to limit than dealing with the disaster now entrained. It's a lot easier to limit the release of toxic substances and greenhouse gases into the environment than to gather them back in once they are recognized as a global threat. And of course it is far more economically sound (and safer) to put money into spreading family planning services and limiting wasteful consumption than dealing with the horrendous costs of overpopulation, overconsumption, and a possible collapse of civilization.[8]

Throughout these pages, we've seen what a difference scientific cultural practices and early intervention can make in preventing later oral-facial problems and the quandary so many adults and parents of older children face in finding effective treatments. How might society be changed so that parents in the future will have superior and more obvious choices when seeking help, and so that their children will be ultimately less likely to need help?

A need among so many for braces is not an inevitable part of growing up, it should now be clear. There are actions you can take on your children's behalf to avoid that need. We hope that now you have also concluded that if the necessity of treatment appears, consideration of orthotropic, forwardontic, or airway-centric orthodontics could be crucial. Finding professionals who will help you avoid the need for braces or give you reliable information on the range of choices and the likely consequences of each is difficult, however. If you have poked your nose into the Internet or your local health community, it is almost certain that you will find little or no access to forwardontic, "orthotropic," or oral-postural therapists.

In medicine and dentistry doctors are obligated to make patients aware of alternative treatments. When there is a viable alternative, the practitioner needs to at least make the patient aware that it is available and explain the pros and cons of each viable choice. And the opportunity of getting a second opinion should always be on the table.

Some of the options that orthodontists provide now include choices such as "if you want your teeth straight we need to pull some of them, and if you want to keep all of them, the alignment may be poor." Surgical options to enlarge jaws so that more teeth will fit in neat rows are also offered, but they naturally are less popular because of pain and expense. However, orthodontists rarely present the option of stimulating jaw growth of young children and sustaining that growth in balance throughout the developmental period. Because this is technically possible, it *should* be presented as an option, despite the greater time, expense, and patient commitment needed. John Mew even resorted to picketing to demand comprehensive informed consent, which means that people, especially those recommended

Image 75.
John Mew
picketing
to demand
orthodontic
reform.

surgery, be made aware of all the alternative treatments available. As he explained, after all, the well-being of our children is at stake.

In a nutshell this is what we believe needs to happen to make appropriate choices much more available and affordable:

First, the public should be informed about the scale of oral-facial problems in society and the existence of preventive methods as well as the full range of potentially curative or at least ameliorative treatments.

Second, health professionals should be educated to communicate thoroughly with patients that have either potential or already existing oral-facial problems on what their treatment choices are.

Third, because one of the main choices should be forwardontics, if only because it is the sole branch that gives full consideration to the effects of treatments on the airway, steps must be taken to greatly increase availability of properly trained practitioners. This means that institutions, especially medical and dental schools, would have to include this type of training in their curricula. All dentists should be aware of forwardontics, issues with resting oral posture, the fundamentals of evolutionary biology, and a critical understanding of the interplay of genetics and culture in oral-facial health.[9]

Dentists should be trained as—and consider themselves to be—oral physicians, whose responsibilities extend beyond the arrangement and health of teeth. This means, among other things, evaluating oral posture and sleep patterns of their patients. Dentists should work together more with medical doctors, especially ear, nose, and throat specialists and sleep physicians, to integrate knowledge in the respective areas to better serve their patients. Just consider, for example, the many dimensions of the problems resulting from mouth breathing. Needless to say pediatricians and family physicians should be made more aware of the problems of oral-facial health, so as to be better equipped to spot problems in their young patients early, and where necessary refer them to specialists.

The shortage of Forwardontists and the training of new ones

A major challenge in adequately addressing oral-facial health is to increase the number of qualified forwardontists and give more attention to what underlies the forwardontic perspective. We can learn from the history of braces. Thirty years ago Sandra's mother in Mexico City used to point at children wearing braces and say, "Look, she is wearing a Volkswagen in her mouth," because orthodontics was an expensive luxury procedure. Today all kinds of braces, even invisible and removable ones, are readily available to practically every middle-class person in industrialized society. More people started paying attention to tooth crowding and its potential negative repercussions, and this precipitated changes in the industry, especially ones that lowered prices. As a result, orthodontics has gained widespread acceptance in the last several decades. Now an analogous cultural evolutionary shift in the direction of forwardontics and a focus on oral posture is needed.

Health professionals in our societies are trained in university programs in medicine and dentistry, and this is where the first cultural change should occur. Clinical university programs need to implement serious training in forwardontics that parallels what is available today for orthodontic specialists. Following the current model for training medical/dental practitioners, residents would start new cases under the guidance of more than one forwardontist on staff. On graduation, the new incoming residents could take over these patients at different stages of treatment. They would also continue to start and supervise new cases. Basically, standard orthodontics should always include forwardontics.

Such changes would take down some of barriers currently keeping forwardontics from becoming commonplace. Because the technique is not taught now in a formal environment, and it is too involved for most private clinicians to master on their own, there are few practitioners of the therapy. Weekend hotel courses are not enough to learn forwardontics. Short "mini-residencies" that are not hands-on are the only current options for learning forwardontics. Because orthotropics entails growth guidance, and growth itself is a slow process, they are not very satisfactory. It takes a practitioner a long time to see a patient from beginning to end of a treatment course because sessions are long, frequent, and intensive compared with those involved in fitting appliances and periodically checking them. If the treatment is solely dependent on one practitioner, it involves a great commitment of time and a heavy responsibility. That, in turn, necessitates a high fee. In the current situation it is emotionally exhausting and usually impractical for solo practitioners to become highly skilled and to continue delivering the service to many patients over decades. The paucity of orthodontic practitioners is so great at present that one woman of our acquaintance had so much trouble getting competent help for her two daughters' serious mouth-breathing problems in the United States that she moved her family to London where the girls could be successfully treated by John and Mike Mew.

In the future, either institutions or group private practices could provide this type of comprehensive care, and the disadvantages of the higher fees would be compensated because forwardontics delivers cures. However, the time commitment and other difficulties of forwardontic treatment underline the great need to change in parallel common cultural practices in society so as to *prevent* most malocclusion and related health problems from ever occurring in the first place.

Prevention

If forwardontic ideas took hold, day care and kindergarten teachers, in support both of practitioners and families, could be enlisted to dedicate some time to GOPex eating, counting, and reading. This would help generate a positive shift in cultural norms, eventually making everybody familiar with oral-facial health problems and their prevention. An added benefit is that the traditionally inspired GOPex practice could help with reading, speaking, and communicating better, giving children more confidence and encouraging them to become more conscious about how they present to the world.

The need for prevention rather than cure is often given lip service, but because of vested interests it is too little acted on. Prevention of some of the diseases that afflict our children and ourselves is what this book is primarily discussing: how we eat and good oral posture are all about prevention. As neurologist Dr. David Perlmutter, author of *The Grain Brain*, put it: "We want to focus not on calling all the king's horses and all the king's men, but on coaxing Humpty Dumpty down from the wall before disaster strikes."[10] For the time being, prevention family by family is likely to be the principal cure to the "how we eat and how we rest" dilemma. Not a solution, but at least a start.

Of course, what may be the most obvious prevention step is the least likely to be taken at present: a return to past "hows" of eating, or at least a modern version of them. Ditch the fast food and canned soups, throw away forks, spoons, and chopsticks, let the ice cream melt, return emphasis on breasts to food sources rather than entertainment devices, close the baby food factories, bring strong chewing to the top of Miss Manners's recom-

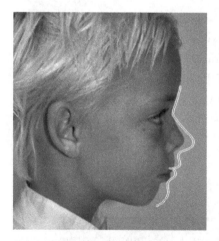 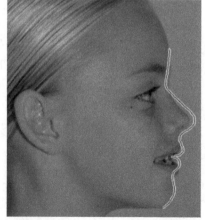

Image 76. This young girl shows the effects of correcting oral posture as well as her overall posture, (Courtesy of Marvin Van Der Linde.)

mendations—and totally disrupt society. Just picture Dad letting Junior pick up a half-chicken or a standing rib roast and letting Junior rip off chunks with his jaws or saw off chunks with his knife while clenching them in his teeth.

Moving in that direction has been discussed,[11] as have been its problems.[12] It clearly isn't going to happen. But anyone can see some partial steps that might help restore average jaw and facial growth to its traditional trajectory. Young kids can't produce bite-sized bits of meat on their own, but Mom can dice tough cuts and watch junior chew. The same can be said for many veggies, and even when children are old enough to use utensils many can be served raw or undercooked. We clearly need to give more thought to how and what we feed our children.

The Need for Cultural Transformation

Most people in industrial societies today of course do not recognize the facial distortion that is common, the spreading of sleep apnea, the extent of children wearing braces, or other symptoms of an epidemic in oral-facial health that can be traced to *how* most people eat and hold their mouths at rest, and other contributing features of humanity's new industrial environment. People tend to accept the world in which they grew up as the standard. But what is common is not necessarily "normal" or healthy. There is growing evidence that a substantial portion of the population *could* have better lives when it comes to oral-facial health and the many consequences of its lack. Indeed, *if spreading knowledge of how to eat could become a civilization-wide priority,* a huge dent could be made in the swelling epidemic of sleep apnea (and possibly a lesser dent in several other nasty diseases ranging from heart problems to mental decline). Many children and their families could, in a more supportive environment, avoid the medical consequences of poor oral posture and the high cost of correcting it entirely.

One obvious need in moving in this direction is for more information to be gathered on the actual scale of the malocclusion problem. We are convinced it is very serious, but that can be considered an informed guess based on fragmentary evidence. Some serious government polling could supply important ammunition to those advocating a change in the norms of how we eat and often breathe. The scale of one aspect of the problem can be seen just in the numbers of children who can't sleep well. Consider the concern expressed by sleep specialist Evelyne Touchette of the University of Quebec at Three Rivers:

Large epidemiological studies carried out in Australia, the United States, Italy and Israel have found that about 30% of preschool children suffer from sleep problems. Persistent sleep problems can affect several aspects of child development (physical, cognitive, social), and can have negative consequences on the early parent–child relationship. It is therefore imperative to identify the factors likely to foster or to hinder good sleep so that childhood sleep problems can be treated.[13]

Sleep disturbance in children, which can include undiagnosed obstructive sleep apnea, has deleterious influences on oral-facial development, as we've seen.[14] But the causes of childhood sleeplessness seem to vary greatly, and its proportional contribution to deviant oral-facial development is not well understood. As one study with which we heartily concur put it:

Oral health problems have risk factors in common with a number of important chronic diseases and conditions such as cardiovascular disease, cancers and injuries. It is wasteful to target each disease separately when they have similar origins. Therefore a strong reason for alliances with other sectors involved in health promotion is to avoid duplication, increase effectiveness and efficiency and reduce isolation by systematically sharing information. Another reason is that the populations with the greatest burdens of all diseases are the deprived and socially excluded.[15]

Changes in laws can also be helpful in fostering cultural evolution even if, as in the case of antismoking ordinances, they are initially not terribly popular. Disapproval of smoking has gradually become widespread in our society. In the case of breastfeeding, passage of state and federal laws making it legal for women to breastfeed in public throughout the United States,[16] have increased the practice, especially among minority women.[17] Large-scale cultural changes are notoriously difficult to accomplish, but the eventual success of campaigns like those for desegregation, gay rights, and the reduction of smoking shows they can work.

With public education efforts and political action (such as campaigning for education programs during school board elections), support for the importance of good oral posture (including classroom chairs that encouraged it[18]) could become part of the curriculum in the first few years of school.[19] Efforts could be made to get more school cafeterias to serve more nutritious and chewy food and schools to offer courses for all students to build knowledge of sound nutrition and proper oral posture. The issues we discuss here could be integrated into health and athletic activities, though it is critical to recognize race and class economic differences in doing so. For example, soft fast food makes up a larger portion of the diets of poor and minority children in comparison with privileged groups.[20] Indeed, the widely accepted erroneous idea that obesity is rooted in a lack of self-control hinders efforts to take social actions (for example, imposing taxes on super-sugared food, controls on serving junk food in school cafeterias, subsidizing stores in poor areas to carry more unprocessed food) that would help prevent obesity.[21]

Human beings are naturally imitative and live in communities, which means that social learning—acquiring knowledge through simple observation and instruction—can also help with the remediation of oral posture problems. Because human beings learn very well by imitating each other, we hope that a self-propagating wave of postural reform might be generated. Consider some major characteristics of today's society. We have become a cyberculture, constantly talking to each other on cell phones, taking pictures of ourselves and our food and posting these, and even playing endless hours of videogames on our phones or iPads. This cyberculture is facing a "perfect storm" of environmental and social problems all competing for solutions. A wave starting with some small changes in the relationship of young children to eating and oral posture and a return to some desirable past practices could contribute to solving some of the most serious of those problems, ones in the areas of oral-facial health and its repercussions. Kids can easily grasp and propagate the simple, fundamental ideas: it's not just "you are what you eat," but also "you are how you eat it and how you rest your mouth."

Gaming, too, could be engaged to improve society. Games might be developed that could help change public knowledge of, and attitudes toward, the health-appearance nexus. They might be able to help stimulate the long-term attention that can make a forwardontic program work. And why not take advantage of the amount of time children engage in virtual activities by incorporating techniques and designs to counteract postural problems, such as the slouching that many people engage in while using cyberdevices?

What we can do as a society is an obvious challenge, but equally important is what each of us can do personally. For this and the other social-environmental problems we have discussed, one can always get involved and be an activist. But much more than dealing with, say, pressuring politicians to deal with climate disruption, you can also take direct action in some cases and make big positive changes in a child's life, both among your own children and grandchildren and by talking with friends, neighbors, and acquaintances who have young children about the issues we've raised.

Governance that treats all people with respect and pays close attention to public health and environmental sustainability, including both *what* people eat and *how* they eat it, might be able to create the institutions necessary to reduce the "unequal access" element in the food crisis, that people suffer and starve because they are poor—and improve everyone's health and nutritional status. A culture rich in "social capital" is one with many cooperative social networks, norms of reciprocity, time to get to know neighbors and interact with them, high levels of trust, high participation in voluntary organizations, and relatively little inequity. In such a society, information on oral-facial health developed in popular literature—health newsletters, magazines, websites, and so on—would spread easily.

People would become more aware of things like mouth breathing, receding chins, stuffy noses, puffy cheeks, and how movie stars swallow. Support groups for those undergoing tedious treatment would be easy to find and join. People would not need to travel far and expensively to get appropriate care; food security and eating well could become the norm. There is evidence that already social networks can help to determine whether people become obese or lose weight[22] and that pledges of behavior[23] and observations of sickness in relatives and friends[24] can influence health status.

There is thus reason to believe that, if there were sufficient social capital, the behavior of parents training their children to eat and rest appropriately could spread from friend to friend and create an epidemic of easy breathing and better health.

Other eating problems in the oral-health area, such as artificial infant-feeding habits, inappropriate weaning diets, limited nonprocessed dietary choices, and exposure to allergens and pollution-loaded enclosed spaces[25] also are widespread, largely preventable, and mostly not dealt with. These, combined with the modern global spread of synthetic chemical poisons (including hormone mimics that may be dangerous in tiny doses) in the air we breathe, the water we drink, and the food we eat, may altogether be contributing to a cascade of food-related health-appearance problems. Included, as we have seen, *could* be contributors to the incidence of heart disease, asthma, attention deficit hyperactivity disorder (ADHD), sleep apnea, sexual dysfunction, mental deficits, and other diseases. Even small contributions to the prevalence of some of these diseases would be, in aggregate, a significant and growing social problem.

In the absence of concerted action, these problems are likely to get worse. On a world scale, one of the factors that will obviously influence *how* we eat is exactly what foods people have access to. Over 2 billion people will be added to the human population by 2050—more people than existed on the entire planet in 1930—most of them in poor countries. There are many hazards, ranging from climate disruption, depletion of groundwater, and soil erosion to the spread of toxic chemicals, extinction of pollinators, and loss of natural pest controls, that will make producing adequate and high-quality food supplies for this giant population expansion difficult.[26]

With so many more people on the planet, demanding more calories and animal protein,[27] existing supply problems will likely increase. And so likely will the diseases associated with the oral-facial health, as huge numbers of additional people are urbanized, relegated to small indoor spaces, and exposed to the industrial way of eating.

A number of people already understand various aspects of the "food problem," starting with the basics of agriculture and nutrition, including what to eat. Now it is necessary to understand the importance of *how* we eat, how we breathe, where we breathe, and how we rest our mouths. These, you now realize, affect such things as the shapes of our children's faces, whether they will need to wear corrective appliances, whether they'll sleep soundly as adults, how long and how well they are likely to live, and indeed even the pace at which they will live their lives. Concerted action is clearly needed to spread awareness of the hidden epidemic, just as it is required if the other problems of the "human predicament" are to be solved. All of us can help to generate such action.

NOTES

INTRODUCTION

1. W. Proffit, H. J. Fields, and L. Moray. 1998. Prevalence of malocclusion and orthodontic treatment need in the United States: Estimates from the NHANES III survey. *Int J Adult Orthodon Orthognath Surg.* 13: 97–106.

2. E. Josefsson, K. Bjerklin, and R. Lindsten. 2007. Malocclusion frequency in Swedish and immigrant adolescents—influence of origin on orthodontic treatment need. *The European Journal of Orthodontics* 29: 79–87.

3. In his lecture "The melting face"; retrieved on February 20, 2016, from www.you tube.com/watch?v=NvoX_wEtwDk.

4. Guilleminault and R. Pelayo. 1998. Sleep-disordered breathing in children. *Annals of Medicine* 30: 350–356.

5. R. A. Settipane. 1999. Complications of allergic rhinitis. *Allergy and Asthma Proceedings*: 209–213.

6. C. All royalties from *Jaws* will go to supporting work related to the subject of the book, making human lives better in a rapidly changing environment.

7. P. Gopalakrishnan and T. Tak. 2011. Obstructive sleep apnea and cardiovascular disease. *Cardiology in Review* 19: 279–290; M. Kohler, J. Pepperell, B. Casadei, S. Craig, N. Crosthwaite, J. Stradling, and R. Davies. 2008b. CPAP and measures of cardiovascular risk in males with OSAS. *European Respiratory Journal* 32: 1488–1496; H. K. Yaggi, J. Concato, W. N. Kernan, J. H. Lichtman, L. M. Brass, and V. Mohsenin. 2005. Obstructive sleep apnea as a risk factor for stroke and death. *New England Journal of Medicine* 353: 2034–2041.

8. J. I. Silverberg and P. Greenland. 2015. Eczema and cardiovascular risk factors in 2 US adult population studies. *Journal of Allergy and Clinical Immunology* 135: 721–728. e726.

9. A. Qureshi, R. D. Ballard, and H. S. Nelson. 2003. Obstructive sleep apnea. *Journal of Allergy and Clinical Immunology* 112: 643–651 A. Sheiham. 2005. Oral health, general health and quality of life. *Bulletin of the World Health Organization* 83: 644–644; A. Sheiham and R. G. Watt. 2000. The common risk factor approach: A rational basis for promoting oral health. *Community Dentistry and Oral Epidemiology* 28: 399–406; R. G. Watt and A. Sheiham. 2012. Integrating the common risk factor approach into a social determinants framework. *Community Dentistry and Oral Epidemiology* 40: 289–296; and Matthew

Walker. 2017. Sleep the good sleep: The role of sleep in causing Alzheimer's disease is undeniable; here's how you can protect yourself. *New Scientist* October 14–20: 30–33.

10. Y. K. Peker, J. Hedner, J. Norum, H. Kraiczi, and J. Carlson. 2002. Increased incidence of cardiovascular disease in middle-aged men with obstructive sleep apnea: A 7-year follow-up. *American Journal of Respiratory and Critical Care Medicine* 166: 159–165.

11. Y. Peker, J. Carlson, and J. Hedner. 2006. Increased incidence of coronary artery disease in sleep apnoea: A long-term follow-up. *European Respiratory Journal* 28: 596–602.

12. A. Qureshi, R. D. Ballard, and H. S. Nelson. 2003. Obstructive sleep apnea. *Journal of Allergy and Clinical Immunology* 112: 643–651.

13. G. Andreou, F. Vlachos, and K. Makanikas. 2014. Effects of chronic obstructive pulmonary disease and obstructive sleep apnea on cognitive functions: Evidence for a common nature. *Sleep Disorders* 2014.

14. Retrieved on February 2, 2016, from http://bit.ly/1OFUnjm.

15. Kirsi Pirilä-Parkkinen, Pertti Pirttiniemi, Peter Nieminen, Heikki Löppönen, Uolevi Tolonen, Ritva Uotila, and Jan Huggare. 1999. Cervical headgear therapy as a factor in obstructive sleep apnea syndrome. *Pediatric Dentistry* 21: 39–45.

16. A.Gibbons. 2014. An evolutionary theory of dentistry. *Science* 336:973–975; J. C. Rose and R. D. Roblee. 2009. Origins of dental crowding and malocclusions: An anthropological perspective. *Compendium of Continuing Education in Dentistry* 30: 292–300.

17. Ron Pinhasi, Vered Eshed, and N. von Cramon-Taubadel. 2015. Incongruity between affinity patterns based on mandibular and lower dental dimensions following the transition to agriculture in the Near East, Anatolia and Europe. *PLoS ONE* 10:e0117301. doi:0117310.0111371/.

18. C. S. Larsen. 2006. The agricultural revolution as environmental catastrophe: Implications for health and lifestyle in the Holocene. *Quaternary International* 150: 12–20.

19. Y. Chida, M. Hamer, J. Wardle, and A. Steptoe. 2008. Do stress-related psychosocial factors contribute to cancer incidence and survival? *Nature Clinical Practice Oncology* 5: 466–475.

20. F. Silva and O. Dutra. 2010. Secular trend in malocclusions. *Orthod Sci Pract* 3: 159–164.

21. M. P. Villa, E. Bernkopf, J. Pagani, V. Broia, M. Montesano, and R. Ronchetti. 2002. Randomized controlled study of an oral jaw-positioning appliance for the treatment of obstructive sleep apnea in children with malocclusion. *American Journal of Respiratory and Critical Care Medicine* 165: 123–127.

CHAPTER 1

1. R. S. Corruccini, G. C. Townsend, L. C. Richards, and T. Brown. 1990. Genetic and environmental determinants of dental occlusal variation in twins of different nationalities. *Human Biology*: 353–367.

2. H. Huggins. 1981. *Why raise ugly kids?* Westport, CT: Arlington House.

3. S. Kahn and S. Wong. 2016. *GOPex: Good Oral Posture Exercises*. Self-published.

4. G. Catlin G. 1861. *Shut Your Mouth and Save Your Life* (original title: *The Breath of Life*). Wiley. (Kindle location 94)

5. Ibid. (Kindle location 83–92)

6. Gapminder. Is child mortality falling? Retrieved on October 26, 2017, from http://bit.ly/1YkmSJc; http://bit.ly/1UsQfDE; http://bit.ly/1U767kT.

7. Anders Olsonn, 2015, Shut your mouth and save your life. Conscious Breathing Available at http://bit.ly/1tjm0sp.

8. G. Catlin. 1861 *Shut Your Mouth and Save Your Life* (original title: *The Breath of Life*). Wiley. (Kindle location 210)

9. Ibid.

10. Ibid.

11. Ibid. (Kindle location 806)

12. Anders Olsonn, 2015, Shut your mouth and save your life. *Conscious Breathing* Available at http://bit.ly/1sA7JaA.

13. J. Goldsmith and S. Stool. 1994. George Catlin's concepts on mouth breathing as presented by Dr. Edbard H. Angle. *Angle Orthodont.* 64: 75–78.

14. Peter W. Lucas, Kai Yang Ang, Zhongquan Sui, Kalpana R. Agrawal, Jonathan F. Prinz, and N. J. Dominy. 2006. A brief review of the recent evolution of the human mouth in physiological and nutritional contexts. *Physiology & Behavior* 89: 36–38.

15. S. Harmand, J. E. Lewis, C. S. Feibel, C. J. Lepre, S. Prat, A. Lenoble, X. Boës, R. L. Quinn, M. Brenet, and A. Arroyo. 2015. 3.3–million-year-old stone tools from Lomekwi 3, West Turkana, Kenya. *Nature* 521: 310–315.

16. Personal communication. August 10, 2015.

17. D. *Lieberman. 2013. The Story of the Human Body: Evolution, Health and Disease.* Penguin UK. (Kindle location 5176)

18. O. Mockers, M. Aubry, and B. Mafart. 2004. Dental crowding in a prehistoric population. *The European Journal of Orthodontics* 26: 151–156.

19. R. Sarig, V. Slon, J. Abbas, H. May, N. Shpack, A. Vardimon, and I. Hershkovitz. 2013. Malocclusion in early anatomically modern human: A reflection on the etiology of modern dental misalignment. *PLoS ONE* 8: DOI: 10.1371/journal.pone.0080771.

20. D. Normando, J. Faber, J. F. Guerreiro, and C. C. A. Quintão. 2011. Dental occlusion in a split Amazon indigenous population: Genetics prevails over environment. *PLoS ONE* 6: e28387.

21. J. P. Evensen and B. Øgaard. 2007. Are malocclusions more prevalent and severe now? A comparative study of medieval skulls from Norway. *American Journal of Orthodontics and Dentofacial Orthopedics* 131: 710–716; J. C. Rose and R. D. Roblee. 2009. Origins of dental crowding and malocclusions: An anthropological perspective. *Compendium of Continuing Education in Dentistry* 30: 292–300; and R. S. Corruccini and E. Pacciani. 1989. "Orthodontistry" and dental occlusion in Etruscans. *The Angle Orthodontist* 59: 61–64.

22. J. P. Evensen and B. Øgaard. 2007. Are malocclusions more prevalent and severe now? A comparative study of medieval skulls from Norway. *American Journal of* Orthodontics and Dentofacial Orthopedics 131: 710–716.

23. B. Mohlin, S. Sagne, and B. Thilander. 1978. The frequency of malocclusion and the craniofacial morphology in a medieval population in Southern Sweden. *Ossa* 5: 57–84.

24. L. Lysell. 1958. A biometric study of occlusion and dental arches in a series of medieval skulls from northern Sweden. *Acta Odontologica Scandinavica* 16: 177–203.

25. C. L. Lavelle. 1972. A comparison between the mandibles of Romano-British and nineteenth century periods. *American Journal of Physical Anthropology* 36: 213–219.

26. C. Harper. 1994. A comparison of medieval and modern dentitions. *The European Journal of Orthodontics* 16: 163–173: and C. L. Lavelle. 1972. A comparison between the mandibles of Romano-British and nineteenth century periods. *American Journal of Physical Anthropology* 36: 213–219.

27. Robert S. Corruccini. 1984. An epidemiologic transition in dental occlusion in world populations. *Amer. J. Orthod.* 86: 419–426; and F; Weiland, E. Jonke, and H. Bantleon. 1997. Secular trends in malocclusion in Austrian men. *The European Journal of Orthodontics* 19: 355–359.

28. Ibid.

29. S. Jew, S. S. AbuMweis, and P. J. Jones. 2009. Evolution of the human diet: Linking our ancestral diet to modern functional foods as a means of chronic disease prevention. *Journal of Medicinal Food* 12: 925–934; and A. Winson. 2013. *The industrial diet: The degradation of food and the struggle for healthy eating.* NYU Press.

30. R. S. Corruccini, G. C. Townsend, L. C. Richards, and T. Brown. 1990. Genetic and environmental determinants of dental occlusal variation in twins of different nationalities. *Human Biology:* 353–367; B. Kawala, J. Antoszewska, and A. Nęcka. 2007. Genetics or environment? A twin-method study of malocclusions. *World Journal of orthodontics* 8; F. Weiland, E. Jonke, and H. Bantleon. 1997. Secular trends in malocclusion in Austrian men. *The European Journal of Orthodontics* 19: 355–359; and E. Defraia, M. Camporesi, A. Marinelli, amd I. Tollaro I. 2008. Morphometric investigation in the skulls of young adults: A comparative study between 19th century and modern Italian samples. *The Angle Orthodontist* 78: 641–646.

31. P. W. Lucas. 2006. Facial dwarfing and dental crowding in relation to diet: 74–82. International Congress Series: Elsevier.

32. F. Silva and O. Dutra. 2010. Secular trend in malocclusions. *Orthod Sci Pract* 3: 159–164.

33. C. S. Larsen. 1995. Biological changes in human populations with agriculture. *Annual Review of Anthropology:* 185–213.

34. Raymond P. Howe, James A. McNamara, and K. A. O'Connor. 1983 An examination of dental crowding and its relationship to tooth size and arch dimension. *American Journal of Orthodontics* 83: 363–373; and F. Silva and O. Dutra. 2010. Secular trend in malocclusions. *Orthod Sci Pract* 3: 159–164.

35. J. W. Friedman. 2007. The prophylactic extraction of third molars: A public health hazard. *American Journal of Public Health* 97: 1554–1559; J. W. Friedman. 2008. Friedman responds. *American Journal of Public Health* 98: 582; and M. E. Nunn, M. D. Fish, R. I. Garcia, E. K. Kaye, R. Figueroa, A. Gohel, M. Ito, H. J. Lee, D, E, Williams, and T. Miyamoto. 2013. Retained asymptomatic third molars and risk for second molar pathology. *Journal of Dental Research* 92: 1095–1099.

CHAPTER 2

1. B. Hockett and J. Haws. 2003. Nutritional ecology and diachronic trends in Paleolithic diet and health. *Evolutionary Anthropology: Issues, News, and Reviews* 12: 211–216.

2. S. B. Eaton and M. Konner. 1985. Paleolithic nutrition: A consideration of its na-

ture and current implications. *New England Journal of Medicine* 312: 283–289; and C. S. Larsen. 2006. The agricultural revolution as environmental catastrophe: Implications for health and lifestyle in the Holocene. *Quaternary International* 150: 12–20.

3. C. J. Ingram, C. A. Mulcare, Y. Itan, M. G. Thomas, and D. M. Swallow. 2009. Lactose digestion and the evolutionary genetics of lactase persistence. *Human Genetics* 124: 579–591.

4. L. A. Frassetto, M. Schloetter, M. Mietus-Synder, R. Morris, and A. Sebastian. 2009. Metabolic and physiologic improvements from consuming a Paleolithic, hunter-gatherer type diet. *European Journal of Clinical Nutrition* 63: 947–955; T. Jönsson, B. Ahrén, G. Pacini, F. Sundler, N. Wierup, S. Steen, T. Sjöberg, M. Ugander, J. Frostegård, and L. Göransson. 2006. A Paleolithic diet confers higher insulin sensitivity, lower C-reactive protein and lower blood pressure than a cereal-based diet in domestic pigs. *Nutrition & Metabolism* 3: 1; T. Jönsson, Y. Granfeldt, B. Ahrén, U,-C, Branell, G. Pålsson, A. Hansson, M. Söderström, and S. Lindeberg S. 2009. Beneficial effects of a Paleolithic diet on cardiovascular risk factors in type 2 diabetes: A randomized cross-over pilot study. *Cardiovasc Diabetol* 8:1–14; M. Österdahl, T. Kocturk, A. Koochek, and P. Wändell. 2008. Effects of a short-term intervention with a Paleolithic diet in healthy volunteers. *European Journal of Clinical Nutrition* 62: 682–685.

5. D. Goose. 1962. Reduction of palate size in modern populations. *Archives of Oral Biology* 7: 343–IN321; Y. Kaifu. 2000. Temporal changes in corpus thickness of the Japanese mandibles. *Bull Natl Sci Mus Ser D* 26: 39–44; C. L. Lavelle. 1972. A comparison between the mandibles of Romano-British and nineteenth century periods. *American Journal of Physical Anthropology* 36: 213–219; D. E. Lieberman, G. E. Krovitz, F. W. Yates, M. Devlin, and M. S. Claire. 2004. Effects of food processing on masticatory strain and craniofacial growth in a retrognathic face. *Journal of Human Evolution* 46: 655–677.

6. C. L. Brace. 1986. Egg on the face, f in the mouth, and the overbite. *American Anthropologist* 88: 695–697.

7. Q. E. Wang. 2015. Chopsticks. Cambridge, UK: Cambridge University Press.

8. C. L. Brace. 1977. Occlusion to the anthropological eye. In *The Biology of Occlusal Development*, J. A. McNamara, ed.: 179–209. Center for Human Growth and Development.

9. G. Catlin. 1861 *Shut Your Mouth and Save Your Life* (original title: *The Breath of Life*). Wiley.

10. D. Lieberman. 2013. *The Story of the Human Body: Evolution, Health and Disease.* Penguin UK. (Kindle Locations 5194–5195).

11. Bucknell University, Roman Food Facts and Worksheets. Retrieved on October 28, 2017, from http://bit.ly/291Jj0J.

12. Sweets throughout Middle Age Europe and the Middle East. Retrieved on October 28, 2017, from http://bit.ly/28YCOL1.

13. Wikipedia. Ice cream. Retrieved on October 28, 2017, from http://bit.ly/2946CKF.

14. Lynne Olver. 2015. Food Timeline FAQS: Baby food. Retrieved on October 28, 2017, from http://bit.ly/292HXnw.

15. F. M. Pottenger. 1946. The effect of heat-processed foods and metabolized vitamin D milk on the dentofacial structures of experimental animals. *American Journal of Orthodontics and Oral Surgery* 32: A467–A485.

16. M. Francis and J. Pottenger. 2012 (1983). *Pottenger's cats: A study in nutrition,* 2nd ed. Price-Pottenger Nutrition Foundation.

17. Beyoindvegetarianism, Lesson of the Pottenger's Cats experiment: Cats are not humans. Retrieved on October 28, 2017, from http://bit.ly/1UNGTVI.

18. W.A. Price. 1939 (2003). *Nutrition and Physical Degeneration.* Price-Pottenger Nutrition Foundation.

19. W. A. Price. 1939 (2003). *Nutrition and physical degeneration.* Price-Pottenger Nutrition Foundation.

20. Daniel Lieberman. 2013. *The story of the human body: Evolution, health, and disease* (Kindle Locations 5179–5181). Knopf Doubleday Publishing Group. Kindle Edition.

21. R. S. Corruccini. 1999. *How anthropology informs the orthodontic diagnosis of malocclusion's causes.* Edwin Mellen Press.

22. W. R. Proffit. 1975. Muscle pressures and tooth position: North American whites and Australian Aborigines. *The Angle Orthodontist* 45: 1–11; and Robert S. Corruccini. 1984. An epidemiologic transition in dental occlusion in world populations. *Amer. J. Orthod.* 86: 419–426.

23. R. Corruccini, A, Henderson, and S. Kaul. 1985. Bite-force variation related to occlusal variation in rural and urban Punjabis (North India). *Archives of Oral Biology* 30: 65–69.

24. P. R. Begg. 1954. Stone Age man's dentition: With reference to anatomically correct occlusion, the etiology of malocclusion, and a technique for its treatment. *American Journal of Orthodontics* 40: 298–312.

25. R. S. Corruccini. 1990. Australian Aboriginal tooth succession, interproximal attrition, and Begg's theory. *American Journal of Orthodontics and Dentofacial Orthopedics* 97: 349–357; M. V. Teja and T. S. Teja. 2013. Anthropology and its relation to orthodontics: Part 2. *APOS Trends in Orthodontics* 3: 45.

26. R. S. Corruccini and R. M. Beecher. 1982. Occlusal variation related to soft diet in a nonhuman primate. *Science* 218: 74–76.

27. J. C. Rose and R. D. Roblee. 2009. Origins of dental crowding and malocclusions: An anthropological perspective. *Compendium of Continuing Education in Dentistry* 30: 292–300.

28. R. S. Corruccini. 1990. Australian Aboriginal tooth succession, interproximal attrition, and Begg's theory. *American Journal of Orthodontics and Dentofacial Orthopedics* 97: 349–357.

29. L. T. Humphrey, I. D. Groote, J. Morales, N. Bartone, S. Collcutt, C. B. Ramsey, and Abdeljalil Bouzouggarh. Earliest evidence for caries and exploitation of starchy plant foods in Pleistocene hunter-gatherers from Morocco. *Proc Natl Acad Sci USA* 111: 954–959.

CHAPTER 3

1. J. M. Diamond. 1989. The great leap forward. *Discover* 10: 50–60.

2. P. R. Ehrlich. 2000. *Human natures: Genes, cultures, and the human prospect.* Island Press.

3. D. E. Lieberman. 2011. *The Evolution of the Human Head.* Harvard University Press.

4. For a recent summary, see Eirik Garnas. 2016. How the Western diet has changed

the human face. *Darwinian Medicine*, February 16. Retrieved on October 28, 2017, from http://bit.ly/24Bjjkv.

5. R. M. Beecher and R. S. Corruccini. 1981. Effects of dietary consistency on craniofacial and occlusal development in the rat. *The Angle Orthodontist* 51: 61–69; and S. A. S. Moimaz, A, J, Í. Garbin, A, M, C, Lima, L, F, Lolli, O, Saliba, and C. A. S. Garbin. 2014. Longitudinal study of habits leading to malocclusion development in childhood. *BMC Oral Health* 14: 96.

6. Robert S. Corruccini. 1984. An epidemiologic transition in dental occlusion in world populations. *Amer. J. Orthod.* 86: 419–426.

7. W. Rock, A, Sabieha, and R. Evans. 2006. A cephalometric comparison of skulls from the fourteenth, sixteenth and twentieth centuries. *British Dental Journal* 200: 33–37; and D. Lieberman. 2013. The story of the human body: Evolution, health and disease. Penguin UK.

8. R, Corruccini, A. Henderson, and S. Kaul. 1985. Bite-force variation related to occlusal variation in rural and urban Punjabis (North India). *Archives of Oral Biology* 30: 65–69; and H. Olasoji and S. Odusanya. 2000. Comparative study of third molar impaction in rural and urban areas of southwestern Nigeria. Tropical Dental Journal: 25–28.

9. R. S. Corruccini and R. M. Beecher. 1982. Occlusal variation related to soft diet in a nonhuman primate. *Science* 218: 74–76; and D. E. Lieberman, G. E. Krovitz, F. W. Yates, M. Devlin, and M. S. Claire. 2004. Effects of food processing on masticatory strain and craniofacial growth in a retrognathic face. *Journal of Human Evolution* 46: 655–677.

10. Environmental Health Perspectives. Retrieved on October 3, 2017, from http://ehp.niehs.nih.gov/120–a402b/.

11. B. Solow, S. Siersbæk-Nielsen, and E. Greve. 1984. Airway adequacy, head posture, and craniofacial morphology. *American Journal of Orthodontics* 86: 214–223.

12. R. Dales, L. Liu, and A. J. Wheeler . 2008. Quality of indoor residential air and health. *Canadian Medical Association Journal* 179:147–152.

13. D. Rosenstreich et al. 1997. The role of cockroach allergy and exposure to cockroach allergen in causing morbidity among inner city children with asthma. *New England Journal of Medicine* 336: 1356–1363.

14. Beate Jacob, Beate Ritz, Ulrike Gehring, Andrea Koch, Wolfgang Bischof, H. E. Wichmann, and J. Heinrich. 2002. Indoor exposure to molds and allergic sensitization. *Environ. Health Perspect.* 110: 647–653; R. E. Dales, H. Zwanenburg, R, Burnett, and C. A. Franklin. 1991. Respiratory health effects of home dampness and molds among Canadian children. *American Journal of Epidemiology* 134: 196–203.

15. T. Husman. 1996. Health effects of indoor-air microorganisms. *Scandinavian Journal of Work, Environment & Health* 22: 5–13.

16. Kathleen Belanger, W. Beckett, E. Triche, M. B. Bracken, T. Holford, P. Ren, J.-E. McSharry, D. R. Gold, T. A. E. Platts-Mills, and B. P. Leaderer. 2003. Symptoms of wheeze and persistent cough in the first year of life: Associations with indoor allergens, air contaminants, and maternal history of asthma. *American Journal of Epidemiology* 158: 195–292; D. P. Skoner. 2001. Allergic rhinitis: Definition, epidemiology, pathophysiology, detection, and diagnosis. *Journal of Allergy and Clinical Immunology* 108: S2–S8; and T. Sih, and O. Mion. 2010. Allergic rhinitis in the child and associated comorbidities. *Pediatric Allergy and Immunology* 21: e107–e113.

17. E. O. Meltzer, M. S. Blaiss, M. J. Derebery, T. A. Mahr, B. R. Gordon, K. K. Sheth, A. L. Simmons, M. A. Wingertzahn, and J. M. Boyle. 2009. Burden of allergic rhinitis: Results from the Pediatric Allergies in America survey. *Journal of Allergy and Clinical Immunology* 124: S43–S70.

18. J. I. Silverberg, E. L. Simpson, H. G. Durkin, and R. Joks. 2013. Prevalence of allergic disease in foreign-born American children. *JAMA Pediatrics* 167: 554–560.

19. S. A. S. Moimaz, A. J. Í. Garbin, A. M. C. Lima, L. F. Lolli, O. Saliba, and C. A. S. Garbin. 2014. Longitudinal study of habits leading to malocclusion development in childhood. *BMC Oral Health* 14: 96.

20. D. Bresolin, P. A. Shapiro, G. G. Shapiro, M. K. Chapko, and S. Dassel. 1983a. Mouth breathing in allergic children: Its relationship to dentofacial development. *American Journal of Orthodontics* 83: 334–340; P. T. M. Faria, A. C. d'O. Ruellas, M. A. N. Matsumoto, W. T. Anselmo-Lima, and F. C. Pereira. 2002. Dentofacial morphology of mouth breathing children. *Brazilian Dental Journal* 13: 129–132; and B. Q. Souki, G. B. Pimenta, M. Q. Souki, L. P. Franco, H. M.. Becker, and J. A. Pinto. 2009. Prevalence of malocclusion among mouth breathing children: Do expectations meet reality? *International Journal of Pediatric Otorhinolaryngology* 73: 767–773.

21. R. R. Abreu, R. L. Rocha, J. A. Lamounier, and Â. F. M. Guerra. 2008. Etiology, clinical manifestations and concurrent findings in mouth-breathing children. *Jornal de pediatria* 84: 529–535; D. Bresolin, P. A. Shapiro, G. G. Shapiro, M. K. Chapko, and S. Dassel. 1983a. Mouth breathing in allergic children: Its relationship to dentofacial development. *American Journal of Orthodontics* 83: 334–340; C. C. Daigle, D. C. Chalupa, F. R. Gibb, P. E. Morrow, G. Oberdörster, M. J. Utell, and M. W. Frampton. 2003. Ultrafine particle deposition in humans during rest and exercise. *Inhalation Toxicology* 15: 539–552; P. T. M. Faria, A. C. d'O. Ruellas, M. A. N. Matsumoto, W. T. Anselmo-Lima, and F. C. Pereira. 2002. Dentofacial morphology of mouth breathing children. *Brazilian Dental Journal* 13: 129–132; J. Paul and R. S. Nanda. 1973. Effect of mouth breathing on dental occlusion. *The Angle Orthodontist* 43: 201–206; and B. Q. Souki, G. B. Pimenta, M. Q. Souki, L. P. Franco, H. M. Becker, and J. A. Pinto. 2009. Prevalence of malocclusion among mouth breathing children: Do expectations meet reality? *International Journal of Pediatric Otorhinolaryngology* 73: 767–773.

22. B. Solow, S. Siersbæk-Nielsen, and E. Greve. 1984. Airway adequacy, head posture, and craniofacial morphology. *American Journal of Orthodontics* 86: 214–223.

23. Ala Al Ali, Stephen Richmond, Hashmat Popat, Rebecca Playle, Timothy Pickles, Alexei I Zhurov, David Marshall, Paul L Rosin, John Henderson, and K. Bonuck. 2015. The influence of snoring, mouth breathing and apnoea on facial morphology in late childhood: A three-dimensional study. *British MedIcal Journal Open* 5: doi:10.1136/bmjopen-2015–009027; and M. B. Marks. 1965. Allergy in relation to orofacial dental deformities in children: A review. *Journal of Allergy* 36: 293–302.

24. K. Behlfelt, S. Linder-Aronson, J. McWilliam, P. Neander, and J. Laage-Hellman. 1990. Cranio-facial morphology in children with and without enlarged tonsils. *The European Journal of Orthodontics* 12: 233–243; S. Linder-Aronson. 1974. Effects of adenoidectomy on dentition and nasopharynx. *American Journal of Orthodontics* 65: 1–15; and D. G. Woodside, S. Linder-Aronson , A. Lundström , and J. McWilliam. 1991. Mandibular and maxillary growth after changed mode of breathing. *American Journal of Orthodontics and Dentofacial Orthopedics* 100: 1–18.

25. S. H. Lee, J. H. Choi, C. Shin, H. M. Lee, S. Y. Kwon, and S. H. Lee. 2007. How does open-mouth breathing influence upper airway anatomy? *Laryngoscope* 117: 1102–1106; and Y. Jefferson. 2010. Mouth breathing: Adverse effects on facial growth, health, academics, and behavior. *Gen. Dent.* 58: 18–25.

26. F. T. Orji, D. K. Adiele, N. G. Umedum, J. O. Akpeh, V. C. Ofoegbu, and J. N. Nwosu. 2016. The clinical and radiological predictors of pulmonary hypertension in children with adenotonsillar hypertrophy. *European Archives of Oto-Rhino-Laryngology*: 1–7.

27. K. Emerich and A. Wojtaszek-Slominska. 2010. Clinical practice. *European Journal of Pediatrics* 169: 651–655.

28. They also have not changed the course of human evolution by a process of natural selection by making people with some hereditary endowments significantly out-reproduce those with others. We don't know how much the problems have influenced reproductive success, and in any case there has been too little time for significant genetic change. Evolution occurs on a time scale of many generations; for people it takes at least thousands of years to generate important genetic changes. Possession of fancy automobiles by the rich cannot yet have significantly altered genetic endowments involved in sexual signaling or planning mating behavior—although the cultural advances in advertising may well have done so. P. R. Ehrlich. 2000. *Human natures: Genes, cultures, and the human prospect.* Island Press.

29. S. J. Olshansky, D. J. Passaro, R. C. Hershow, J. Layden, B. A. Carnes, J. Brody, L. Hayflick, R. N. Butler, D. B. Allison, and D. S. Ludwig. 2005. A potential decline in life expectancy in the United States in the 21st century. *New England Journal of Medicine* 352: 1138–1145.

CHAPTER 4

1. Population Reference Bureau. 2016. *2016 World Population Data Sheet.* Population Reference Bureau.

2. J. R. C. Mew. 2004a. The postural basis of malocclusion: A philosophical overview. *The American Journal of Orthodontics and Dentofacial Orthopedics* 126: 729–738.

3. H. Valladas, J. Clottes, J.-M. Geneste, M. A. Garcia, M. Arnold, H. Cachier, and N. Tisnérat-Laborde. 2001. Palaeolithic paintings: Evolution of prehistoric cave art. *Nature* 413: 479–479.

4. D. J. Lewis-Williams and J. Clottes J. 1998. The mind in the cave—The cave in the mind: Altered consciousness in the Upper Paleolithic. *Anthropology of Consciousness* 9: 13–21; and D. S. Whitley. 2009. *Cave paintings and the human spirit: The origin of creativity and belief.* Prometheus Books.

5. A. Bouzouggar, N. Barton, M. Vanhaeren, F. d'Errico, S. Collcutt, T. Higham, E. Hodge, S. Parfitt, E. Rhodes, and J.-L. Schwenninger. 2007. 82,000–year-old shell beads from North Africa and implications for the origins of modern human behavior. *Proceedings of the National Academy of Sciences* 104: 9964–9969.

6. P. Chin Evans and A. R. McConnell. 2003. Do racial minorities respond in the same way to mainstream beauty standards? Social comparison processes in Asian, black, and white women. *Self and Identity* 2: 153–167.

7. C. C. I. Hall. 1995. Asian eyes: Body image and eating disorders of Asian and Asian American women. *Eating Disorders* 3: 8–19.

8. D. E. Lieberman, G. E. Krovitz, F. W. Yates, M. Devlin, and M. S. Claire. 2004. Effects of food processing on masticatory strain and craniofacial growth in a retrognathic face. *Journal of Human Evolution* 46: 655–677.

9. G. Korkhaus G. 1960. Present orthodontic thought in Germany: jaw widening with active appliances in cases of mouth breathing. *American Journal of Orthodontics* 46:187–206, Mew JRC. 2004a. The postural basis of malocclusion: A philosophical overview. *The American Journal of Orthodontics and Dentofacial Orthopedics* 126:;729–738; P. Defabjanis. 2004. Impact of nasal airway obstruction on dentofacial development and sleep disturbances in children: Preliminary notes. *Journal of Clinical Pediatric Dentistry* 27: 95–100.; and K. Lopatien? and A. Babarskas A. 2002. Malocclusion and upper airway obstruction. *Medicina* 38: 277–283.

10. E. Gokhale, and S. Adams. 2008. 8 steps to a pain-free back. Stanford, CA: Pendo Press; J. Kratěnová, K. ŽEjglicová, and V. Filipová. 2007. Prevalence and risk factors of poor posture in school children in the Czech Republic. *Journal of School Health* 77: 131–137.

11. Ibid.

12. D. Yosifon and P. N. Stearns. 1998. The rise and fall of American posture. *The American Historical Review* 103: 1057–1095.

13. A. T. Masi and J. C. Hannon. 2008. Human resting muscle tone (HRMT): Narrative introduction and modern concepts. *Journal of Bodywork and Movement Therapies* 12: 320–332.

14. K. Grimmer. 1997. An investigation of poor cervical resting posture. *Australian Journal of Physiotherapy* 43: 7–16.

15. P. B. M. Conti, E. Sakano, M. Â. G. d'O. Ribeiro, C. I. S. Schivinski, and J. D. Ribeiro. 2011b. Assessment of the body posture of mouth-breathing children and adolescents. *Jornal de pediatria* 87: 357–363; P. Nicolakis, M. Nicolakis, E. Piehslinger, G. Ebenbichler, M. Vachuda, C. Kirtley, and V. Fialka-Moser. 2000. Relationship between craniomandibular disorders and poor posture. *Cranio: The Journal of Craniomandibular Practice* 18: 106–112; and E. F. Wright, M. A. Domenech, and J. R. Fischer. 2000. Usefulness of posture training for patients with temporomandibular disorders. *The Journal of the American Dental Association* 131: 202–210.

16. H. Nittono, M. Fukushima, A. Yano, and H. Moriya. 2012. The power of kawaii: Viewing cute images promotes a careful behavior and narrows attentional focus. *PLoS ONE* 7: e46362.

17. V. A. De Menezes, R. B. Leal, R. S. Pessoa, and R. M. E. S. Pontes. 2006. Prevalence and factors related to mouth breathing in school children at the Santo Amaro project-Recife, 2005. *Brazilian Journal of Otorhinolaryngology* 72: 394–398.

18. C. Sforza, R. Peretta, G. Grandi, G. Ferronato, and V. F. Ferrario. 2007. Three-dimensional facial morphometry in skeletal Class III patients: A non-invasive study of soft-tissue changes before and after orthognathic surgery. *British Journal of Oral and Maxillofacial Surgery* 45: 138–144.

19. A. A. Ali, S. Richmond, H. Popat, R. Playle, T. Pickles, A. I. Zhurov, D. Marshall, P. L. Rosin, J. Henderson, and K. Bonuck. 2015. The influence of snoring, mouth breathing and apnoea on facial morphology in late childhood: Three-dimensional study. *British Medical Journal* 5: e009027; S. A. Schendel, J. Eisenfeld, W. H. Bell, B. N. Epker, and David J. Mishelevich. 1976. The long face syndrome: Vertical maxillary excess. *American*

Journal of Orthodontics 70: 398–408; and L. P. Tourne. 1990. The long face syndrome and impairment of the nasopharyngeal airway. *Angle Orthod* 60: 167–176.

20. Y. Jefferson. 2004. Facial beauty: Establishing a universal standard. *International Journal of Orthodontics* 15: 9–26.

21. N. J. Pollock. 1995. Cultural elaborations of obesity: Fattening practices in Pacific societies. *Asia Pacific J Clin Nutr* 4: 357–360, ibid.

22. A. Brewis, S. McGarvey, J. Jones, and B. Swinburn B. 1998. Perceptions of body size in Pacific Islanders. *International Journal of Obesity* 22: 185–189.

23. Retrieved on December 13, 2015, from http://bit.ly/1P25zHc.

24. G. Rhodes, S. Yoshikawa, A. Clark, K. Lee, R. McKay, and S. Akamatsu. 2001. Attractiveness of facial averageness and symmetry in non-Western cultures: In search of biologically based standards of beauty. *Perception* 30: 611–625.

25. J. F. Cross and J. Cross. 1971a. Age, sex, race, and the perception of facial beauty. *Developmental Psychology* 5: 433; D. Jones and K. Hill. 1993. Criteria of facial attractiveness in five populations. *Human Nature* 4: 271–296; F. B. Naini, J. P. Moss, and D. S. Gill. 2006. The enigma of facial beauty: Esthetics, proportions, deformity, and controversy. *American Journal of Orthodontics and Dentofacial Orthopedics* 130: 277–282; G. Rhodes. 2006. The evolutionary psychology of facial beauty. *Annu. Rev. Psychol.* 57: 199–226; A. J. Rubenstein, J. H. Langlois, and L. A. Roggman. 2002. What makes a face attractive and why: The role of averageness in defining facial beauty. In *Facial Attractiveness: Evolutionary, Cognitive, and Social Perspectives. Advances in Visual Cognition*, vol. 1, G. Rhodes and L. A. Zebrowitz, eds.: 1–33. Ablex Publishing.

26. N. Barber. 1995. The evolutionary psychology of physical attractiveness: Sexual selection and human morphology. *Ethology and Sociobiology* 16: 395–424; D. M. Buss and M. Barnes. 1986. Preferences in human mate selection. *Journal of Personality and Social Psychology* 50: 559–570; K. Grammer and R. Thornhill. 1994. Human (Homo sapiens) facial attractiveness and sexual selection: The role of symmetry and averageness. *Journal of Comparative Psychology* 108: 233–242; L. Mealey, R. Bridgstock, and G. C. Townsend. 1999. Symmetry and perceived facial attractiveness: A monozygotic co-twin comparison. *Journal of Personality and Social Psychology* 76: 151–158; I. S. Penton-Voak, B. C. Jones, A. C. Little, S. Baker, B. Tiddeman, D. M. Burt, and D. I. Perrett. 2001. Symmetry, sexual dimorphism in facial proportions and male facial attractiveness. *Proc. R. Soc. Lond. B* 258; D. I. Perrett, D. M. Burt, I. S. Penton-Voak, K. J. Lee, D. A. Rowland, and R. Edwards. 1999. Symmetry and human facial attractiveness. *Evolution and human behavior* 20: 295–307; D. I. Perrett, K. J. Lee, I. Penton-Voak, D. Rowland, S. Yoshikawa, D. M. Burt, S. P. Henzi, D. L. Castles, and S. Akamatsu. 1998. Effects of sexual dimorphism on facial attractiveness. *Nature* 394: 884–887; S. C. Roberts, J. Havlicek, J. Flegr, M. Hruskova, A, C, Little, B. C. Jones, D. I. Perrett, and M. Petrie. 2004. Female facial attractiveness increases during the fertile phase of the menstrual cycle. *Proc. R. Soc. Lond. B* 271: S270–S272; R. Thornhill and S. W. Gangestad. 1999. Facial attractiveness. *Trends in Cognitive Sciences* 3: 452–460; and J. S. Winston, J. O'Doherty, J. M. Kilner, D. I. Perrett, and R. J. Dolan. 2007. Brain systems for assessing facial attractiveness. *Neuropsychologia* 45: 195–206.

27. J. Gottschall. 2007. Greater emphasis on female attractiveness in Homo sapiens: A revised solution to an old evolutionary riddle. *Evolutionary Psychology* 5: 147470490700500208.

28. M. Bashour. 2006a. History and current concepts in the analysis of facial attrac-

tiveness. *Plastic and Reconstructive Surgery* 118:741–756; M. Bashour. 2006b. An objective system for measuring facial attractiveness. *Plastic and Reconstructive Surgery* 118: 757–774.

29. C.-C. Carbon, T. Grüter, M. Grüter, J. E. Weber, and A. Lueschow. 2010. Dissociation of facial attractiveness and distinctiveness processing in congenital prosopagnosia. *Visual Cognition* 18: 641–654; K. Nakamura, R. Kawashima, S. Nagumo, K. Ito, M. Sugiura, T. Kato, A. Nakamura, K. Hatano, K. Kubota, and H. Fukuda. 1998. Neuroanatomical correlates of the assessment of facial attractiveness. *Neuroreport* 9: 753–757.

30. S. C. Roberts, J. Havlicek, J. Flegr, M. Hruskova, A. C. Little, B. C. Jones, D. I. Perrett, and M. Petrie. 2004. Female facial attractiveness increases during the fertile phase of the menstrual cycle. *Proc. R. Soc. Lond. B* 271: S270–S272

31. R. Bull and N. Rumsey. 1988. *The social psychology of facial appearance.* New York: Springer-Verlag; F. Conterio and L. L. Cavalli-Sforza. 1960. Selezione per caratteri quantaiativi nell'uomo. *Atti. Ass. Genet. Ital.* 5: 295–304.

32. N. Etcoff. 1999. *Survival of the prettiest: The science of beauty.* Anchor/Doubleday.

33. A. Iglesias-Linares, R.-M. Yáñez-Vico, B. Moreno-Manteca, A. M. Moreno-Fernández, A. Mendoza-Mendoza, and E. Solano-Reina. 2011. Common standards in facial esthetics: Craniofacial analysis of most attractive black and white subjects according to *People* magazine during previous 10 years. *Journal of Oral and Maxillofacial Surgery* 69: e216–e224.

34. R. Bull and N. Rumsey N. 1988. *The social psychology of facial appearance.* Springer-Verlag.

35. L. Lowenstein. 1978. The bullied and non-bullied child. *Bulletin of the British Psychological Society* 31: 316–318.

36. N. Berggren, H. Jordahl, and P. Poutvaara. 2010. The looks of a winner: Beauty and electoral success. *Journal of Public Economics* 94: 8–15; G. Lutz. 2010. The electoral success of beauties and beasts. *Swiss Political Science Review* 16: 457–480; and U. Rosar, M. Klein, and T. Beckers. 2008. The frog pond beauty contest: Physical attractiveness and electoral success of the constituency candidates at the North Rhine-Westphalia state election of 2005. *European Journal of Political Research* 47: 64–79.

37. C. Bosman, G. Pfann, J. Biddle, and D. Hamermesh. 1997. Business success and businesses' beauty capital. *NBER Working Paper* Number 6083; I. H. Frieze and J. E. Olson. 1991. Attractiveness and income for men and women in management. *Journal of Applied Social Psychology* 21: 1039–1057; C. M. Marlowe, S. L. Schneider, and C. E. Nelson. 1996. Gender and attractiveness biases in hiring decisions: Are more experienced managers less biased?. *Journal of Applied Psychology* 81: 11–21; and G. A. Pfann, J. E. Biddle, D. S. Hamermesh, and C. M. Bosman. 2000. Business success and businesses' beauty capital. *Economics Letters* 67: 201–207.

38. B. Fink, N. Neave, J. T. Manning, and K. Grammer. 2006. Facial symmetry and judgements of attractiveness, health and personality. *Personality and Individual Differences* 41: 491–499; F. B. Furlow, T. Armijo-Prewirr, S. W. Gangestad, R. Thornhill. 1997. Fluctuating asymmetry and psychometric intelligence. *Proc. R. Soc. Lond. B* 264: 823–829; K. Grammer, B. Fink, A. P. Møller, and J. T. Manning. 2005. Physical attractiveness and health: Comment on Weeden and Sabini (2005). *Psychological Bulletin* 131: 658–661; J. J. A. Henderson and J. M. Anglin. 2003. Facial attractiveness predicts longevity. *Evolution and Human Behavior* 24: 351–356; D. Umberson and M. Hughes. 1987. The impact

of physical attractiveness on achievement and psychological well-being. *Social Psychology Quarterly* 50: 227–236; J. Weeden and J. Sabini. 2005. Physical attractiveness and health in Western societies: A review. *Psychological Bulletin* 131: 635–653; and D. W. Zaidel, S. M. Aarde, and K. Baig. 2005. Appearance of symmetry, beauty, and health in human faces. *Brain and Cognition* 57: 261–263.

39. J. Stewart. 1980. Defendant's attractiveness as a factor in the outcome of criminal trials: An observational study. *Journal of Applied Social Psychology* 10: 348–361.

40. G. Patzer. 1985. *The physical attractiveness phenomena*. Plenum.

41. J. F. Cross and J. Cross. 1971b. Age, sex, race, and the perception of facial beauty. *Developmental Psychology* 5: 433–439.

42. J. H. Langlois, J. M. Ritter, L. A. Roggman, and L. S. Vaughn. 1991. Facial diversity and infant preferences for attractive faces. *Developmental Psychology* 27: 79–84; and K. Lewis. 1969. Infants responses to facial stimuli during the first year of life. *Developmental Psychology* 1: 75–86.

43. Emma Young. 2016. Who do you think you are? 4 rules can help you. *New Scientist*. January 27. Available at http://bit.ly/1Pios6K.

44. G. Rhodes, L. Jeffery, T. L. Watson, C. W. Clifford, and K. Nakayama. 2003. Fitting the mind to the world face adaptation and attractiveness aftereffects. *Psychological Science* 14: 558–566.

45. S. Strom. 2014. Study examines efficacy of taxes on sugary drinks. *New York Times*, June 2. Available at http://nyti.ms/1gW6H1L.

46. R, N, Proctor. 2011. *Golden holocaust: Origins of the cigarette catastrophe and the case for abolition*. University of California Press.

CHAPTER 5

1. K. L. Boyd. 2011. Darwinian Dentistry Part 1. An Evolutionary Perspective on the Etiology of Malocclusion: 34–40. Available on the website of the American Orthodontic Society at www.orthodontics.com.

2. J. R. C. Mew. 1981. The aetiology of malocclusion: Can the tropic premise assist our understanding? *British Dental Journal* 151: 296–301; J. R. C. Mew. 2004a. The postural basis of malocclusion: A philosophical overview. *The American Journal of Orthodontics and Dentofacial Orthopedics* 126: 729–738.

3. D. Bresolin, P. A. Shapiro, G. G. Shapiro, M. K. Chapko, and S. Dassel. 1983a. Mouth breathing in allergic children: Its relationship to dentofacial development. *American Journal of Orthodontics* 83: 334–340; A. Hannuksela. 1981. The effect of moderate and severe atopy on the facial skeleton. *The European Journal of Orthodontics* 3: 187–19; C. Oulis, G. Vadiakas, J. Ekonomides, and J. Dratsa. 1993. The effect of hypertrophic adenoids and tonsils on the development of posterior crossbite and oral habits. *The Journal of Clinical Pediatric Dentistry* 18: 197–201; and G. M. Trask, G. G. Shapiro, and P. A. Shapiro. 1987. The effects of perennial allergic rhinitis on dental and skeletal development: A comparison of sibling pairs. *American Journal of Orthodontics and Dentofacial Orthopedics* 92: 286–293.

4. M. B. Marks. 1965. Allergy in relation to orofacial dental deformities in children: A review. *Journal of Allergy* 36: 293–302.

5. S. Linder-Aronson, D. Woodside, and A. Lundströ. 1986. Mandibular growth direction following adenoidectomy. *American Journal of Orthodontics* 89: 273–284.

6. P. R. Ehrlich and A. H. Ehrlich. 2009. *The dominant animal: Human evolution and the environment,* 2nd edition. Island Press.

7. P. Lieberman. 2007. Evolution of human language. *Current Anthropology* 48: 39–66.

8. J. M. Diamond. 1989. The great leap forward. *Discover* 10: 50–60; J. M. Diamond. 1991. *The rise and fall of the third chimpanzee.* Radius.

9. T. M. Davidson. 2003. The great leap gorward: The anatomic basis for the acquisition of speech and obstructive sleep apnea. *Sleep Medicine* 4: 185–194.

10. Ibid.

11. Ibid.

12. S. Baldrigui, A. Pinzan, C. Zwicker, C. Michelini, D. Barros, and F. Elias. 2001. The importance of the natural milk to prevent myofuncional and orthodontics alterations. *Rev Dent Press Ortodon Ortop Facial* 6: 111–121; S. A. S. Moimaz, A. J. Í. Garbin, A. M. C. Lima, L. F. Lolli, O. Saliba, and C. A. S. Garbin. 2014. Longitudinal study of habits leading to malocclusion development in childhood. *BMC Oral Health* 14: 96.

13. S. Baldrigui, A. Pinzan, C. Zwicker, C. Michelini, D. Barros, and F. Elias. 2001. The importance of the natural milk to prevent myofuncional and orthodontics alterations. *Rev Dent Press Ortodon Ortop Facial* 6: 111–121; G. Carvalho. 1998. Amamentação é prevenção das alterações funcionais e estruturais do sistema estomatognático. *Odontologia Ensino e Pesquisa, Cruzeiro* 2: 39–48; C. M. M. Gimenez, A. B. Ad. Moraes, A. P. Bertoz, F. A. Bertoz, and G. B. Ambrosano. 2008. First childhood malocclusion's prevalence and its relation with breast feeding and oral habits. *Revista* Dental Press de Ortodontia e Ortopedia Facial 13: 70–83.

14. K. G. Peres, A. J. Barros, M. A. Peres, and C. G. Victora. 2007. Effects of breast-feeding and sucking habits on malocclusion in a birth cohort study. *Revista de saude Publica* 41: 343–350.

15. S. Sexton and R. Natale. 2009. Risks and benefits of pacifiers. *American Family Physician* 79.

16. D. Viggiano, D. Fasano, G. Monaco, and L. Strohmenger. 2004. Breast feeding, bottle feeding, and non-nutritive sucking: Effects on occlusion in deciduous dentition. *Archives of Disease in Childhood* 89: 1121–1123.

17. Kevin Boyd video, Industrialization and Crooked Teeth. Retrieved on October 28, 2017, from http://bit.ly/1QAX8RR.

18. O. Silva Filho, A. Cavassan, M. Rego, and P. Silva. 2003. Sucking habits and malocclusion: Epidemiology in deciduous dentition. *Rev Clin Ortodontia Dental Press* 2: 57–74; and D. Viggiano, D. Fasano, G. Monaco, and L. Strohmenger. 2004. Breast feeding, bottle feeding, and non-nutritive sucking; Effects on occlusion in deciduous dentition. *Archives of Disease in Childhood* 89: 1121–1123.

19. D. Lieberman. 2013. The story of the human body: Evolution, health and disease. Penguin UK.

20. C. Safina. 2015. Beyond words: What animals think and feel. Henry Holt.

21. D. Bresolin, P. A. Shapiro, G. G. Shapiro, M. K. Chapko, and S. Dassel. 1983a. Mouth breathing in allergic children: Its relationship to dentofacial development. *American Journal of Orthodontics* 83: 334–340; D. Bresolin, P. A. Sharpiro, G. G. Shapiro, M. K. Chapko, and S. Dassel. 1983b. Mouth breathing in allergic children: Its relationship to dentofacial development. *American Journal of Orthodontics and Dentofacial Orthopedics* 83:

334–339; P. T. M. Faria, A. C. d'O. Ruellas, M. A. N. Matsumoto, W. T. Anselmo-Lima, and F. C. Pereira. 2002. Dentofacial morphology of mouth breathing children. *Brazilian Dental Journal* 13: 129–132; Y. Jefferson. 2010. Mouth breathing: Adverse effects on facial growth, health, academics, and behavior. *Gen. Dent.* 58: 18–25; S. H. Lee, J. H. Choi, C. Shin, H. M. Lee, S. Y. Kwon, and S. H. Lee. 2007. How does open-mouth breathing influence upper airway anatomy? *Laryngoscope* 117: 1102–1106; S. E. Mattar, W. Anselmo-Lima, F. Valera, and M. Matsumoto. 2004a. Skeletal and occlusal characteristics in mouth-breathing pre-school children. *Journal of Clinical Pediatric Dentistry* 28: 315–318; P. D. Neiva, R, N, Kirkwood, and R. Godinho. 2009. Orientation and position of head posture, scapula and thoracic spine in mouth-breathing children. *International Journal of Pediatric Otorhinolaryngology* 73: 227–236; and B. Q. Souki, G. B. Pimenta, M. Q. Souki, L. P. Franco, H. M. Becker, and J. A. Pinto. 2009. Prevalence of malocclusion among mouth breathing children: Do expectations meet reality? *International Journal of Pediatric Otorhinolaryngology* 73: 767–773.

22. Y. Jefferson. 2010. Mouth breathing: Adverse effects on facial growth, health, academics, and behavior. *Gen. Dent.* 58: 18–25; P. Defabjanis. 2004. Impact of nasal airway obstruction on dentofacial development and sleep disturbances in children: Preliminary notes. *Journal of Clinical Pediatric Dentistry* 27: 95–100; S. Raskin, M. Limme, and R. Poirrier 2000. [Could mouth breathing lead to obstructive sleep apnea syndromes? A preliminary study]. *L'Orthodontie francaise* 71: 27–35.

23. E. P. Harvold, B. S. Tomer, K. Vargervik, and G. Chierici. 1981. Primate experiments on oral respiration. *Am J Orthod.* 79: 159–172.

24. Ibid.

25. E. P. Harvold. 1968. The role of function in the etiology and treatment of malocclusion. *American Journal of Orthodontics* 54: 883–896.

26. A. A. Ali, S. Richmond, H. Popat, R. Playle, T. Pickles, A. I. Zhurov, D. Marshall, P. L. Rosin, J. Henderson, and K. Bonuck. 2015. The influence of snoring, mouth breathing and apnoea on facial morphology in late childhood: Three-dimensional study. *British Medical Journal* 5: e009027; and L. P. Tourne. 1990. The long face syndrome and impairment of the nasopharyngeal airway. *Angle Orthod* 60: 167–176.

27. D. Johnston, O. Hunt, C. Johnston, D. Burden, M. Stevenson, and P. Hepper. 2005. The influence of lower face vertical proportion on facial attractiveness. *The European Journal of Orthodontics* 27: 349–354; C. Sforza, A. Peretta, G. Grandi, G. Ferronato, and V. F. Ferrario. 2007. Three-dimensional facial morphometry in skeletal Class III patients: A non-invasive study of soft-tissue changes before and after orthognathic surgery. *British Journal of Oral and Maxillofacial Surgery* 45: 138–144.

28. J. D. Rugh and C. J. Drago. 1981. Vertical dimension: A study of clinical rest position and jaw muscle activity. *The Journal of Prosthetic Dentistry* 45: 670–675.

29. M. B. Marks. 1965. Allergy in relation to orofacial dental deformities in children: A review. *Journal of Allergy* 36: 293–302.

30. P. S. Bergeson and J. C. Shaw. 2001. Are infants really obligatory nasal breathers? *Clin Pediatr* 40: 567–569.

31. S. Linder-Aronson. 1970. Adenoids: Their effect on mode of breathing and nasal airflow and their relationship to characteristics of the facial skeleton and the dentition. *Acta Otolaryngol. Suppl.* 265: 1–132.

32. B. Schaub, R. Lauener, and E. von Mutius. 2006. The many faces of the hygiene hypothesis. *Journal of Allergy and Clinical Immunology* 117: 969–977.

33. S. A. S. Moimaz, A. J. Í. Garbin, A. M. C. Lima, L. F. Lolli, O. Saliba, and C. A. S. Garbin. 2014. Longitudinal study of habits leading to malocclusion development in childhood. *BMC Oral Health* 14: 96.

34. D, W, Sellen. 2007. Evolution of infant and young child feeding: Implications for contemporary public health. *Annu. Rev. Nutr.* 27: 123–148.

35. A. Patki. 2007. Eat dirt and avoid atopy: The hygiene hypothesis revisited. *Indian Journal of Dermatology, Venereology, and Leprology 73: 2*.

36. M. Garrett, M. Hooper, B. Hooper, P. Rayment, and M. Abramson. 1999. Increased risk of allergy in children due to formaldehyde exposure in homes. *Allergy* 54: 330–337.

37. P. Vedanthan, P. Mahesh, R. Vedanthan, A. Holla, and L. Ah. 2006. Effect of animal contact and microbial exposures on the prevalence of atopy and asthma in urban vs rural children in India. *Ann Allergy Asthma Immunol.* 96: 571–578.

38. R. Rafael. 1990. Nasopharyngeal obstruction as a cause of malocclusion [in Spanish]. *Pract Odontol.* 11: 11–15, 17, 19–20 passim; and R. A. Settipane. 1999. Complications of allergic rhinitis. *Allergy and Asthma Proceedings*: 209–213.

39. R. A. Settipane. 1999. Complications of allergic rhinitis. *Allergy and Asthma Proceedings*: 209–213; and T. A. Platts-Mills. 2007. The role of indoor allergens in chronic allergic disease. *Journal of Allergy and Clinical Immunology* 119: 297. Available at http://dailym.ai/21wgZw7.

40. G. Gallerano, G. Ruoppolo, and A. Silvestri. 2012. Myofunctional and speech rehabilitation after orthodontic-surgical treatment of dento-maxillofacial dysgnathia. *Progress in Orthodontics* 13: 57–68.

41. S. W. Herring. 1993. Formation of the vertebrate face epigenetic and functional influences. *American Zoologist* 33: 472–483; J. Varrela. 1990. Genetic and epigenetic regulation of craniofacial development. *Proceedings of the Finnish Dental Society. Suomen Hammaslaakariseuran toimituksia* 87: 239–244; T. F. Schilling and P. V. Thorogood. 2000. Development and evolution of the vertebrate skull. *Linnean Society Symposium Series*: 57–84; T. E. Parsons, E. J. Schmidt, J. C. Boughner, H. A. Jamniczky, R. S. Marcucio, and B. Hallgrímsson. 2011. Epigenetic integration of the developing brain and face. *Developmental Dynamics* 240: 2233–2244; and K. M. Xiong, R. E. Peterson, and W. Heideman. 2008. Aryl hydrocarbon receptor-mediated down-regulation of sox9b causes jaw malformation in zebrafish embryos. *Molecular Pharmacology* 74: 1544–1553.

42. C. Ackroyd, N. K. Humphrey, and E. K. Warrington. 1974. Lasting effects of early blindness: A case study. *Quarterly Journal of Experimental Psychology* 26: 114–124; and S. Carlson and L. E. A. Hyvärinen. 1983. Visual rehabilitation after long lasting early blindness. *Acta Ophthalmologica* 61: 701–713.

43. E, Huber, J. M. Webster, A. A. Brewer, D. I. A. MacLeod, B. A. Wandell, G. M. Boynton, A, R, Wade, and I. Fine. 2015. A lack of experience-dependent plasticity after more than a decade of recovered sight. *Psychological Science* 26: 393–401.

44. Y. Ostrovsky, A. Andalman, and P. Sinha. 2006. Vision following extended congenital blindness. *Psychological Science* 17: 1009–1014.

45. J. S. Johnson and E. L. Newport. 1989. Critical period effects in second language

learning: The influence of maturational state on the acquisition of English as a second language. *Cognitive Psychology* 21: 60–99.

46. R. M. DeKeyser. 2000. The robustness of critical period effects in second language acquisition. *Studies in Second Language Acquisition* 22: 499–533.

47. J. R. C. Mew. 2013. *The cause and cure of malocclusion*. Self-published.

48. C. M. M. Gimenez, A. B. Ad. Moraes, A. P. Bertoz, F. A. Bertoz, and G. B. Ambrosano. 2008. First childhood malocclusion's prevalence and its relation with breast feeding and oral habits. *Revista Dental Press de Ortodontia e Ortopedia Facial* 13: 70–83.

49. C. Paschetta, S. de Azevedo, L. Castillo, N. Martínez-Abadías, M. Hernández, D. E. Lieberman, and R. González-José. 2010. The influence of masticatory loading on craniofacial morphology: A test case across technological transitions in the Ohio Valley. *American Journal of Physical Anthropology* 141: 297–314; Ron Pinhasi, Vered Eshed, and N. Cramon-Taubadel. 2015. Incongruity between affinity patterns based on mandibular and lower dental dimensions following the transition to agriculture in the Near East, Anatolia and Europe. *PLoS ONE* 10: e0117301. doi:0117310.0111371/; P. W. Lucas. 2006. Facial dwarfing and dental crowding in relation to diet. *International Congress Series*: 74–82. Elsevier; and N. von Cramon-Taubadel. 2011. Global human mandibular variation reflects differences in agricultural and hunter-gatherer subsistence strategies. *Proceedings of the National Academy of Sciences* 108: 19546–19551.

50. V. Eshed, A. Gopher, and I. Hershkovitz. 2006. Tooth wear and dental pathology at the advent of agriculture: New evidence from the Levant. *American Journal of Physical Anthropology* 130: 145–159.

51. C. Dürrwächter, O. E. Craig, M. J. Collins, J. Burger, and K. W. Alt. 2006. Beyond the grave: Variability in Neolithic diets in Southern Germany? *Journal of Archaeological Science* 33: 39–48; C. Paschetta, S. de Azevedo, L. Castillo, N. Martínez-Abadías, M. Hernández, D. E. Lieberman, and R. González-José. 2010. The influence of masticatory loading on craniofacial morphology: A test case across technological transitions in the Ohio Valley. *American Journal of Physical Anthropology* 141: 297–314; and M. Richards. 2002. A brief review of the archaeological evidence for Palaeolithic and Neolithic subsistence. *European Journal of Clinical Nutrition* 56: 16.

52. C. S. Larsen. 1995. Biological changes in human populations with agriculture. *Annual Review of Anthropology*: 185–213; and M. N. Cohen and G. M. M. Crane-Kramer. 2007. Ancient health: Skeletal indicators of agricultural and economic intensification. University Press of Florida.

53. A. Crompton and P. Parker. 1978. Evolution of the mammalian masticatory apparatus: The fossil record shows how mammals evolved both complex chewing mechanisms and an effective middle ear, two structures that distinguish them from reptiles. *American Scientist* 66: 192–201; M. J. Ravosa. 1996. Jaw morphology and function in living and fossil Old World monkeys. *International Journal of Primatology* 17: 909–932; and C. F. Ross, D. A. Reed, R. L. Washington, A. Eckhardt, F. Anapol, and N. Shahnoor. 2009. Scaling of chew cycle duration in primates. *American Journal of Physical Anthropology* 138: 30–44.

54. D. Bresolin, G. G. Shapiro, P. A. Shapiro, S. Dassel, C. T. Furukawa, et al. 1984. Facial characteristics of children who breathe through the mouth. *Pediatrics* 73: 622–625; D. Bresolin, P. A. Sharpiro, G. G. Shapiro, M. K. Chapko, and S. Dassel. 1983b. Mouth breathing in allergic children: Its relationship to dentofacial development. *American Journal*

of Orthodontics and Dentofacial Orthopedics 83: 334–339; J. R. C. Mew. 2004a. The postural basis of malocclusion: A philosophical overview. *The American Journal of Orthodontics and Dentofacial Orthopedics* 126: 729–738; J. B. Palmer and K. M. Hiiemae. 2003. Eating and breathing: Interactions between respiration and feeding of solid food. *Dysphagia* 18: 169–178; W. A. Price. 1939 (2003). *Nutrition and physical degeneration.* Price-Pottenger Nutrition Foundation; E. Townsend and N. J. Pitchford. 2012. Baby knows best? The impact of weaning style on food preferences and body mass index in early childhood in a case-controlled sample. *BMJ Open* 2: e000298. doi:000210.001136/bmjopen-002011–000298; and H. Yamaguchi and K. Sueishi. 2003. Malocclusion associated with abnormal posture. *Bull. Tokyo Dent. Coll.* 44: 43–54.

55. F. Neiva, D. Cattoni, J. Ramos, and H. Issler. 2003. Early weaning: Implications to oral motor development. *J Pediatr (Rio J)* 79: 7–12.

56. P. Gluckman and M. Hanson. 2007. Developmental plasticity and human disease: Research directions. *Journal of Internal Medicine* 261: 461–471.

57. C. L. Lavelle. 1972. A comparison between the mandibles of Romano-British and nineteenth century periods. *American Journal of Physical Anthropology* 36: 213–219; W. Rock, A. Sabieha and R. Evans. 2006. A cephalometric comparison of skulls from the fourteenth, sixteenth and twentieth centuries. *British Dental Journal* 200: 33–37; and J. P. Evensen and B. Øgaard. 2007. Are malocclusions more prevalent and severe now? A comparative study of medieval skulls from Norway. *American Journal of Orthodontics and Dentofacial Orthopedics* 131: 710–716.

58. Y. Takahashi, D. M. Kipnis, W. H. Daughaday. 1968. Growth hormone secretion during sleep. *The Journal of Clinical Investigation* 67: 2079–2090.

59. G. Brandenberger, C. Gronfier, zzzzzzf. Chapotot, C. Simon, and F. Piquard. 2000. Effect of sleep deprivation on overall 24 h growth-hormone secretion. *The Lancet* 356: 1408; Y. Takahashi, D. M. Kipnis, W. H. Daughaday. 1968. Growth hormone secretion during sleep. *The Journal of Clinical Investigation* 67: 2079–2090.

CHAPTER 6

1. B. S. McEwen. 2006. Sleep deprivation as a neurobiologic and physiologic stressor: Allostasis and allostatic load. *Metabolism: Clinical and Experimental* 55: S20–S23.

2. K. J. Reichmuth, D. Austin, J. B. Skatrud, and T. Young. 2005. *Association of sleep apnea and type II diabetes: A population-based study. American Journal of Respiratory and Critical Care Medicine* 172: 1590–1595.

3. M. R. Mannarino, F. Di Filippo, and M. Pirro. 2012. Obstructive sleep apnea syndrome. *European Journal of Internal Medicine* 23: 586–593.

4. A. C. Halbower, M. Degaonkar, P. B. Barker, C. J. Earley, C. L. Marcus, P. L. Smith, M. C. Prahme, and E. M. Mahone. 2006. Childhood obstructive sleep apnea associates with neuropsychological deficits and neuronal brain injury. *PLoS Med* 3: e301; Y.-S. Huang, C.. Guilleminault, H.-Y. Li, C.-M. Yang, Y.-Y. Wu, and N.-H. Chen. 2007. Attention-deficit/hyperactivity disorder with obstructive sleep apnea: A treatment outcome study. *Sleep Medicine* 8: 18–30; K. B. Kim. 2015. How has our interest in the airway changed over 100 years? *American Journal of Orthodontics and Dentofacial Orthopedics* 148: 740–747; S. A. Mulvaney, J. L. Goodwin, W. J. Morgan, G. R. Rosen, S. F. Quan, and K. L. Kaemingk. 2006. Behavior problems associated with sleep disordered breathing in school-

aged children: The Tucson Children's Assessment of Sleep Apnea Study. *Journal of Pediatric Psychology* 31: 322–330; and R. Silvestri, A. Gagliano, I. Aricò, T. Calarese, C. Cedro, O. Bruni, R. Condurso, E. Germanò, G. Gervasi, and R. Siracusano. 2009. Sleep disorders in children with attention-deficit/hyperactivity disorder (ADHD) recorded overnight by video-polysomnography. *Sleep Medicine* 10: 1132–1138.

5. C. Guilleminault and S. Sullivan. 2014. Towards restoration of continuous nasal breathing as the ultimate treatment goal in pediatric obstructive sleep apnea. *Enliven: Pediatr Neonatol Biol* 1: 001.

6. E. Glatz-Noll and R. Berg. 1991. Oral disfunction in children with Down's syndrome: An evaluation of treatment effects by means of video-registration. *Eur. J. Orthod.* 13: 446–451; and S. Linder-Aronson. 1970. Adenoids: Their effect on mode of breathing and nasal airflow and their relationship to characteristics of the facial skeleton and the dentition. *Acta Otolaryngol. Suppl.* 265: 1–132; and J. R. C. Mew. 2004b. The postural basis of malocclusion: A philosophical overview. *The American Journal of Orthodontics and Dentofacial Orthopedics* 126: 729–738.

7. V. A. De Menezes, R. B. Leal, R. S. Pessoa, and R. M. E. S. Pontes. 2006. Prevalence and factors related to mouth breathing in school children at the Santo Amaro project-Recife, 2005. *Brazilian Journal of Otorhinolaryngology* 72: 394–398.

8. P. Vig, D. Sarver, D. Hall, and B. Warren. 1981. Quantitative evaluation of airflow in relation to facial morphology. *Am J Orthod* 79: 272–273.

9. J. R. C. Mew. 2004. The postural basis of malocclusion: A philosophical overview. *The American Journal of Orthodontics and Dentofacial Orthopedics* 126: 729–738.

10. J. R. Harkema, S. A. Carey, and J. G. Wagner. 2006. The nose revisited: A brief review of the comparative structure, function, and toxicologic pathology of the nasal epithelium. *Toxicologic Pathology* 34: 252–269.

11. A. L. C. Foresi, D. Olivieri, and G. Cremona. 2007. Alveolar-derived exhaled nitric oxide is reduced in obstructive sleep apnea syndrome. *Chest* 132; and J. O. N. Lundberg and A. Weitzberg. 1999. Nitric oxide in man. *Thorax* 54: 947–952.

12. M. J. Griffiths and T. W. Evans. 2005. Inhaled nitric oxide therapy in adults. *New England Journal of Medicine* 353: 2683–2695.

13. T. Aznar, A. Galán, I. Marin, and A. Domínguez. 2006. Dental arch diameters and relationships to oral habits. *The Angle Orthodontist* 76: 441–445; and R. A. Settipane. 1999. Complications of allergic rhinitis. *Allergy and Asthma Proceedings*: 209–213.

14. V. A. De Menezes, L. B. Leal, R. S. Pessoa, and R. M. E. S. Pontes. 2006. Prevalence and factors related to mouth breathing in school children at the Santo Amaro project-Recife, 2005. *Brazilian Journal of Otorhinolaryngology* 72: 394–398.

15. Colin Fernandez. 2016. Sleeping with your mouth open damages teeth "as much as a fizzy drink before bed": Dry mouth causes acid levels to rise, eroding teeth. *Daily Mail,* February 5. Retrieved on November 30, 2017, from http://dailym.ai/21wgZw7.

16. E. S. Frenkel and K. Ribbeck. 2015. Salivary mucins protect surfaces from colonization by cariogenic bacteria. *Applied and Environmental Microbiology* 81(1): 332–338.

17. P. McKeown. 2011. *Close your mouth: Self-help Buteyko manual.* Amazon Digital Services.

18. March 2016, p. 8; see also www.statisticbrain.com/sleeping-disorder-statistics/.

19. May 28–June 3, 2016, p. 5.

20. J. E. Remmers. 1990. Sleeping and breathing. *Chest* 97 (suppl): 77S-80S; J. E. Remmers, W. J. DeGroot, E. K. Sauerland, and A. M. Anch. 1978. Pathogenesis of upper airway occlusion during sleep. *J. Appl. Physiol. : Respirat. Environ. Exercise Physiol.* 44: 931–938.

21. Cited by Dr. Bill Hang at the Face Focused Orthodontics. Lecture heard by S. K. at AAPMD (American Association of Physiologic Medicine and Dentistry), Conference in Oakland, CA, 2013.

22. R. Sapolsky. 1998. *Why zebras don't get ulcers: An updated guide to stress, stress-related diseases, and coping.* W. H. Freeman & Co.

23. Personal communication, December 11, 2015

24. Mandy Oaklander. 2015. Lack of sleep dramatically raises your risk for getting sick. *Time*, 31 August.Retrieved on October 28, 2017, from http://ti.me/1JJa8F2.

25. R. M. Sapolsky. 2004. *Why zebras don't get ulcers*, 3rd edition. Henry Holt and Company.

26. S. Loth, B. Petruson, G. Lindstedt, et al. 1998. Improved nasal breathing in snorers increases nocturnal growth hormone secretion and serum concentrations of insulin-like growth factor. *Rhinology* 36: 179–183.

27. D. Gozal, F. Hakim, and L. Kheirandish-Gozal. 2013. Chemoreceptors, baroreceptors, and autonomic deregulation in children with obstructive sleep apnea. *Respiratory Physiology & Neurobiology* 185: 177–185.

28. R. M. Sapolsky. 2004. *Why zebras con't get ulcers*, 3rd edition. Henry Holt and Company.

29. G. Grassi, G. Seravalle, and F. Quarti-Trevano F. 2010. The "neuroadrenergic hypothesis" in hypertension: Current evidence. *Experimental Physiology* 95: 581–586.

30. Steven Reinberg. 2015. Sleep apnea devices lower blood Pressure. HealthDay. December 1. Retrieved on October 28, 2017, from http://bit.ly/263TfSj.

31. W. W. Schmidt-Nowara, D. B. Coultas, C. Wiggins, B. E. Skipper, and J. M. Samet. 1990. Snoring in a Hispanic-American population: Risk factors and association with hypertension and other morbidity. *Archives of Internal Medicine* 150: 597–601.

32. Snoring Statistics. Statistics related to snoring problems, SleepDisordersGuide.com. Retrieved on October 28, 2017, from http://bit.ly/1tBY7Nm.

33. J. Stradling and J. Crosby. 1991. Predictors and prevalence of obstructive sleep apnoea and snoring in 1001 middle aged men. *Thorax* 46: 85–90.

34. Ibid.

35. W, W, Schmidt-Nowara, D, B, Coultas, C. Wiggins, B. E. Skipper, and J. M. Samet. 1990. Snoring in a Hispanic-American population: Risk factors and association with hypertension and other morbidity. *Archives of Internal Medicine* 150: 597–601.

36. D. Gozal. 1998. Sleep-disordered breathing and school performance in children. *Pediatrics* 102: 616.

37. P. Counter and J. A. Wilson. 2004. The management of simple snoring. *Sleep Medicine Reviews* 8: 433–441.

38. M. Kohler, K. Lushington, R. Couper, J. Martin, C. van den Heuvel, Y. Pamula, and D. Kennedy. 2008a. Obesity and risk of sleep related upper airway obstruction in caucasian children. *J Clin Sleep Med* 4: 129–136.

39. C. M. Hill, A. M. Hogan, N. Onugha, D. Harrison, S. Cooper, V. J. McGrigor,

A. Datta, and F. J. Kirkham. 2006. Increased cerebral blood flow velocity in children with mild sleep-disordered breathing: A possible association with abnormal neuropsychological function. *Pediatrics* 118.

40. D. Gozal. 1998. Sleep-disordered breathing and school performance in children. *Pediatrics* 102: 616.

41. M. E. Barnes, E. A. Huss, K. N. Garrod, E. Van Raay, E. Dayyat, D. Gozal, and D. L. Molfese. 2009b. Impairments in attention in occasionally snoring children: An event-related potential study. *Developmental Neuropsychology* 34: 629–649; A. P. F. Key, D. L. Molfese, L. O'Brien, and D. Gozal. 2009. Sleep-disordered breathing affects auditory processing in 5–7-year-old children: Evidence from brain recordings. *Developmental Neuropsychology* 34(5): 615–628; L. M. O'Brien, C. B. Mervis, C. R. Holbrook, J. L. Bruner, C. J. Klaus, J. Rutherford, T. J. Raffield, and D. Gozal. 2004. Neurobehavioral implications of habitual snoring in children. *Pediatrics* 114: 44–49.

42. S. Miano, M. Paolino, R. Peraita-Adrados, M. Montesano, S. Barberi, and M. Villa. 2009. Prevalence of eeg paroxysmal activity in a population of children with obstructive sleep apnea syndrome. *Sleep* 32: 522–529.

43. C. M. Hill, A. M. Hogan, N. Onugha, D. Harrison, S. Cooper, V. J. McGrigor, A. Datta, and F. J. Kirkham. 2006. Increased cerebral blood flow velocity in children with mild sleep-disordered breathing: A possible association with abnormal neuropsychological function. *Pediatrics* 118.

44. T. Young, P. E. Peppard, and D. J. Gottlieb. 2002. Epidemiology of obstructive sleep apnea: A population health perspective. *American Journal of Respiratory and Critical Care Medicine* 165: 1217–1239.

45. T. Peltomäki. 2007. The effect of mode of breathing on craniofacial growth—revisited. *The European Journal of Orthodontics* 29: 426–429.

46. C. Guilleminault and S. Sullivan. 2014. Towards restoration of continuous nasal breathing as the ultimate treatment goal in pediatric obstructive sleep apnea. *Enliven: Pediatr Neonatol Biol* 1: 001.

47. Y. M. Ahn. 2010. Treatment of obstructive sleep apnea in children. *Korean Journal of Pediatrics* 53: 872–879; and J. Chan, J. C. Edman, and Peter J. Koltai. 2004. Obstructive sleep apnea in children. *Am Fam Physician* 69: 1147–1154.

48. Quoted in a lecture by forwardontic dentist William M. Hang.

49. William M. Hang, communication at the AAPMD (American Association of Physiologic Medicine and Dentistry) Conference in Oakland, CA, June 14–15, 2013.

50. Y. M. Betancourt-Fursow de Jiménez, J. C. Jiménez-León, and C. S. Jiménez-Betancourt. 2006. Attention deficit hyperactivity disorder and sleep disorders [Article in Spanish]. *Rev. Neurol.* 13: S37–51; P. B. Conti, E. Sakano, M. Â. Ribeiro, C. I. Schivinski, and J. D. Ribeiro. 2011a. Assessment of the body posture of mouth-breathing children and adolescents. *J. Pediatr (Rio J)* 87: 357–363; M. Hallani, J. R. Wheatley, and T. C. Amis. 2008. Enforced mouth breathing decreases lung function in mild asthmatics. *Respirology* 13: 553–558; Y. Jefferson. 2010. Mouth breathing: Adverse effects on facial growth, health, academics, and behavior. *Gen. Dent.* 58: 18–25; P. K. Mangla and M. P. Menon. 1981. Effect of nasal and oral breathing on exercise-induced asthma. *Clin Allergy* 11: 433–439; S. K. Steinsvåg, B. Skadberg, and K. Bredesen. 2007. Nasal symptoms and signs in children suffering from asthma. *International Journal of Pediatric Otorhinolaryngology* 71: 615–621;

and M. E. Barnes, Elizabeth A. Huss, Krista N. Garrod, Eric Van Raay, Ehab Dayyat, David Gozal, and D. L. Molfese. 2009a. Impairments in attention in occasionally snoring children: An event-related potential study. *Dev Neuropsychol* 34: 629–649.

51. S. H. Sheldon. 2010. Obstructive sleep apnea and bruxism in children. *Sleep Medicine Clinics* 5: 163–168.

52. A. G. Tilkian, C. Guilleminault, J. S. Schroeder, K. L. Lehrman, F. B. Simmons, and W. C. Dement. 1977. Sleep-induced apnea syndrome: Prevalence of cardiac arrhythmias and their reversal after tracheostomy. *The American Journal of Medicine* 63: 348–358.

53. Y. M. Betancourt-Fursow de Jiménez, L. C. Jiménez-León, and C. S. Jiménez-Betancourt. 2006. Attention deficit hyperactivity disorder and sleep disorders [Article in Spanish]. *Rev. Neurol.* 13: S37–51; R. D. Chervin, C. Bassetti, D. A. Ganoczy, and K. J. Pituch. 1997. Pediatrics and sleep symptoms of sleep disorders, inattention, and hyperactivity in children. *Sleep* 20: 1185–1192; T. Fidan, and V. Fidan. 2008. The impact of adenotonsillectomy on attention-deficit hyperactivity and disruptive behavioral symptoms. *The Eurasian Journal of Medicine* 40: 14–17; L. O'Brien et al. 2003. Sleep and neurobehavioral characteristics of 5–to 7–year-old children with parentally reported symptoms of attention-deficit/hyperactivity disorder. *Pediatrics* 111: 554–563; and K. Sedky, D. S. Bennett, and K. S. Carvalho. 2014. Attention deficit hyperactivity disorder and sleep disordered breathing in pediatric populations: A meta-analysis. *Sleep Medicine Reviews* 18: 349e356.

54. D. J. Timms. 1990. Rapid maxillary expansion in the treatment of nocturnal enuresis. *The Angle Orthodontist* 60: 229–233.

55. U. Schültz-Fransson and J. Kurol. 2008. Rapid maxillary expansion effects on nocturnal enuresis in children: A follow-up study. *Angle Orthod.* 78: 201–208.

56. There is speculation that sleep apnea in pregnant women and in their newborns may cause autism, although this is a very contentious subject. See D. E. Wardly. 2014. Autism, sleep disordered breathing, and intracranial hypertension: The circumstantial evidence. *Medical Hypotheses and Research* 9: 1–33.

57. R. M. Sapolsky. 2004. *Why zebras don't get ulcers*. 3rd edition. Henry Holt and Company.

58. M. Butt, G. Dwivedi, O. Khair, and G. Y. Lip. 2010. Obstructive sleep apnea and cardiovascular disease. *International Journal of Cardiology* 139: 7–16; and P. Gopalakrishnan and T. Tak. 2011. Obstructive sleep apnea and cardiovascular disease. *Cardiology in Review* 19: 279–290.

59. C. Xin, W. Zhang, L. Wang, D. Yang, and J. Wang. 2015. Changes of visual field and optic nerve fiber layer in patients with OSAS. *Sleep Breath* 19: 129–134.

60. A. N. Vgontzas, D. A. Papanicolaou, E. O. Bixler, K. Hopper, A. Lotsikas, H.-M. Lin, A. Kales, and G. P. Chrousos. 2000. Sleep apnea and daytime sleepiness and fatigue: Relation to visceral obesity, insulin resistance, and hypercytokinemia. *The Journal of Clinical Endocrinology and Metabolism* 85: 1151–1158.

61. M. A. Daulatzai. 2013. Death by a thousand cuts in Alzheimer's disease: Hypoxia—the prodrome. *Neurotox Res* 24: 216–243; and K. B. Kim. 2015. How has our interest in the airway changed over 100 years? *American Journal of Orthodontics and Dentofacial Orthopedics* 148: 740–747.

62. Mark Wheeler. 2015. UCLA researchers provide first evidence of how obstructive

sleep apnea damages the brain. UCLA Newsroom. September 1. Retrieved on November 22, 2015, from http://bit.ly/1RkngBS.

63. D. W. Beebe and D. Gozal D. 2002. Obstructive sleep apnea and the prefrontal cortex: Towards a comprehensive model linking nocturnal upper airway obstruction to daytime cognitive and behavioral deficits. *Journal of Sleep Research* 11: 1–16; B. Naëgelé, V. Thouvard, J.-L. Pépin, P. Lévy, C. Bonnet, J. E. Perret, J. Pellat, and C. Feuerstein. 1995. Deficits of cognitive executive functions in patients with sleep apnea syndrome. *Sleep: Journal of Sleep Research & Sleep Medicine*; S. K. Rhodes, K. C. Shimoda, L. R. Waid, P. M. O'Neil, M. J. Oexmann, N. A. Collop, and S. M. Willi. 1995. Neurocognitive deficits in morbidly obese children with obstructive sleep apnea. *The Journal of Pediatrics* 127: 741–744; and J. Molano, D. Kleindorfer, L. McClure, F. Unverzagt, V. Wadley, and V. Howard. 2015. The association of sleep apnea and stroke with cognitive performance: The reasons for geographic and racial differences in stroke (REGARDS) study. *Neurology* 84: Supplement S53.005.

64. M.-A, Bédard, J. Montplaisir, F. Richer, I. Rouleau, and J. Malo. 1991. Obstructive sleep apnea syndrome: Pathogenesis of neuropsychological deficits. *Journal of Clinical and Experimental Neuropsychology* 13: 950–964; and S. K. Rhodes, K. C. Shimoda, L. R. Waid, P. M. O'Neil, M. J. Oexmann, N. A. Collop, and S. M. Willi. 1995. Neurocognitive deficits in morbidly obese children with obstructive sleep apnea. *The Journal of Pediatrics* 127: 741–744.

65. M. Alchanatis, N. Zias, N. Deligiorgis, A. Amfilochiou, G. Dionellis, and D. Orphanidou. 2005. Sleep apnea-related cognitive deficits and intelligence: An implication of cognitive reserve theory. *Journal of Sleep Research* 14: 69–75.

66. D. Gozal, F. Hakim, and L. Kheirandish-Gozal. 2013. Chemoreceptors, baroreceptors, and autonomic deregulation in children with obstructive sleep apnea. *Respiratory Physiology & Neurobiology* 185: 177–185.

67. C. M. Hill, A. M. Hogan, N. Onugha, D. Harrison, S. Cooper, V. J. McGrigor, A. Datta, and F. J. Kirkham. 2006. Increased cerebral blood flow velocity in children with mild sleep-disordered breathing: A possible association with abnormal neuropsychological function. *Pediatrics* 118.

68. P. Mehra, M. Downie, M. C. Pita, and L. M. Wolford. 2001. Pharyngeal airway space after counterclockwise rotation of the maxillomandibular complex. *Am J Dentofacial Orthop* 120: 154–159; and N. Powell. 2005. Upper airway surgery does have a major role in the treatment of obstructive sleep apnea: "The tail end of the dog." *Journal of Clinical Sleep Medicine* 1: 236–240.

69. D. Wardly, L. M. Wolford, and V. Veerappan. 2016. Idiopathic intracranial hypertension eliminated by counterclockwise maxillomandibular advancement: A case report. *Cranio: The Journal of Craniomandibular and Sleep Practice* DOI: 10.1080/08869634.2016.1201634.

70. N. Powell. 2005. Upper airway surgery does have a major role in the treatment of obstructive sleep apnea: "The tail end of the dog." *Journal of Clinical Sleep Medicine* 1: 236–240; M. Tselnik and M. Anthony Pogrel. 2000. Assessment of the pharyngeal airway space after mandibular setback surgery. *Journal of Oral and Maxillofacial Surgery* 58: 282–285; M. Kawakami, K. Yamamoto, M. Fujimoto, K. Ohgi, M. Inoue, and T. Kirita. 2005. Changes in tongue and hyoid positions, and posterior airway space following mandibular

setback surgery. *Journal of Cranio-Maxillofacial Surgery* 33: 107–110; .J. C. Quintero and J. McCain J. 2012. Total airway volume increase through OMfS measured with cone beam CT: A case report. September. orthotown.com.

71. K. Degerliyurt, K. Ueki, Y. Hashiba, K. Marukawa, K. Nakagawa, and E. Yamamoto. 2008. A comparative CT evaluation of pharyngeal airway changes in class III patients receiving bimaxillary surgery or mandibular setback surgery. *Oral Surgery, Oral Medicine, Oral Pathology, Oral Radiology and Endodontology* 105: 495–502.

72. R. Wijey. 2014. Orthognathic surgery: The definitive answer? *International Journal of Orthodontics* 25(4): 67–68.

73. C. H. Won, K. K. Li, and C. Guilleminault C. 2008. Surgical treatment of obstructive sleep apnea: Upper airway and maxillomandibular surgery. *Proceedings of the American Thoracic Society* 5: 193–199.

74. L. Ferini-Strambi, C. Baietto, M. Di Gioia, P. Castaldi, C. Castronovo, M. Zucconi, and S. Cappa. 2003. Cognitive dysfunction in patients with obstructive sleep apnea (OSA): Partial reversibility after continuous positive airway pressure (CPAP). *Brain Research Bulletin* 61: 87–92.

75. R. Davies and J. R. Stradling. 1990. The relationship between neck circumference, radiographic pharyngeal anatomy, and the obstructive sleep apnoea syndrome. *Eur Respir J* 3: 509–514; R. J. O. Davies, N. J. Ali, and J. R. Stradling. 1992. Neck circumference and other clinical features in the diagnosis of the obstructive sleep apnoea syndrome. *Thorax* 47: 101–105; and R. J. Schwab, M. Pasirstein, R, Pierson, A. Mackley, R. Hachadoorian, R. Arens, G. Maislin, and A. I. Pack. 2003. Identification of upper airway anatomic risk factors for obstructive sleep apnea with volumetric magnetic resonance imaging. *American Journal of Respiratory and Critical Care Medicine* 168: 222–530.

CHAPTER 7

1. G. Catlin. 1861 *Shut your mouth and save your life* (original title: *The breath of life*). Wiley.

2. K. G. Peres, A. M. Cascaes, M. A. Peres, F. F. Demarco, I. S. Santos, A. Matijasevich, and A. J. Barros. 2015. Exclusive breastfeeding and tisk of dental malocclusion. *Pediatrics* 136 :e60–e67; and S. A. S. Moimaz, A. J. Í. Garbin, A. M. C. Lima, L. F. Lolli, O. Saliba, and C. A. S. Garbin. 2014. Longitudinal study of habits leading to malocclusion development in childhood. *BMC Oral Health* 14: 96.

3. M. S. Fewtrell, J. B. Morgan, C. Duggan, G. Gunnlaugsson, P. L. Hibberd, A. Lucas, and R. E. Kleinman. 2007. Optimal duration of exclusive breastfeeding: What is the evidence to support current recommendations? *The American Journal of Clinical Nutrition* 85: 635S-638S.

4. A. L. García, S. Raza, A. Parrett, and C. M. Wright. 2013. Nutritional content of infant commercial weaning foods in the UK. *Archives of Disease in Childhood* 98: 793–797.

5. Studies of rats have shown that those fed a liquid diet after weaning had changes in their facial bones and jaw musculature; see Z. Liu, K. Ikeda, S. Harada, Y. Kasahara, and G. Ito. 1998. Functional properties of jaw and tongue muscles in rats fed a liquid diet after being weaned. *Journal of Dental Research* 77: 366–376.

6. Website for organization: www.babyledweaning.com/.

7. Email, October 20, 2015.

8. Personal communication with Sandra, London, October 2015.

9. J. Diamond. 2012. *The world until yesterday*. Viking.

10. M. Bergamini, F. Pierleoni, A. Gizdulich, and C. Bergamini. 2008. Dental occlusion and body posture: A surface EMG study. *Cranio* 26: 25–32; S. Kiwamu, R. Mehta Noshir, F. Abdallah Emad, Albert G. Forgione, H. Hiroshi, K. Takao, and Y. Atsuro. 2014. Examination of the relationship between mandibular position and body posture. *Cranio* 25(4): 237–249; and D. Manfredini, T. Castroflorio, G. Perinetti, and L. Guarda-Nardini. 2012. Dental occlusion, body posture and temporomandibular disorders: Where we are now and where we are heading for. *Journal of Oral Rehabilitation* 39: 463–471.

11. M. Rocabado, B. E. Johnston Jr., and M. G. Blakney. 1982. Physical therapy and dentistry: An overview: A perspective. *Journal of Craniomandibular Practice* 1: 46–49; and B. Solow and L. Sonnesen. 1998. Head posture and malocclusions. *The European Journal of Orthodontics* 20: 685–693.

12. E. Antunovic. 2008. Strollers, baby carriers, and infant stress: Horizontal versus upright transport in early infancy. Retrieved on December 20, 2015, from http://bit.ly/1ZpXyR3.

13. M. C. Frank, K. Simmons, D. Yurovsky, and G. Pusiol. 2013. Developmental and postural changes in children's visual access to faces. *Proceedings of the 35th Annual Meeting of the Cognitive Science Society, Austin, TX*: 454–459.

14. S. Zeedyk. 2008. What's life in a baby buggy like? The impact of buggy orientation on parent–infant interaction and infant stress. London: National Literacy Trust. Retrieved pn November 21, 2008, from www.suttontrust.com/research-paper/whats-life-baby-buggy-like-impact-buggy-orientation-parent-infant-interaction-infant-stress/.

15. J. R. Harkema, S. A. Carey, and J. G. Wagner. 2006. The nose revisited: A brief review of the comparative structure, function, and toxicologic pathology of the nasal epithelium. *Toxicologic pathology* 34: 252–269.

16. Personal communication, February 11, 2016.

17. R. Dales, L. Liu, and A. J. Wheeler. 2008. Quality of indoor residential air and health. *Canadian Medical Association Journal* 179: 147–152.

18. J. M. Samet, M. C. Marbury, and J. D. Spengler. 1988. Health effects and sources of indoor air pollution. Part II. *American Review of Respiratory Disease* 137: 221–242.

19. M. Garrett, M. Hooper, B. Hooper, P. Rayment, and M. Abramson. 1999. Increased risk of allergy in children due to formaldehyde exposure in homes. *Allergy* 54: 330–337.

20. J. L. Sublet, J. Seltzer, R. Burkhead, P. B. Williams, H. J. Wedner, and W. Phipatanakul. 2010. Air filters and air cleaners: Rostrum by the American Academy of Allergy, Asthma & Immunology Indoor Allergen Committee. *Journal of Allergy and Clinical Immunology* 125: 32–38.

21. L. Roberts, W. Smith, L. Jorm, M. Patel, R. M. Douglas, and C, McGilchrist. 2000. Effect of infection control measures on the frequency of upper respiratory infection in child care: A randomized, controlled trial. *Pediatrics* 105: 738–742.

22. C. Guilleminault and S. Sullivan S. 2014. Towards restoration of continuous nasal breathing as the ultimate treatment goal in pediatric obstructive sleep apnea. *Enliven: Pediatr Neonatol Biol* 1: 001.

23. P. McKeown. 2011. *Close your mouth: Self-help Buteyko manual*. Amazon Digital Services; S. Cooper, J. Oborne, S. Newton, V. Harrison, J. T. Coon, S. Lewis, and A. Tatters-

field. 2003. Effect of two breathing exercises (Buteyko and pranayama) in asthma: A randomised controlled trial. *Thorax* 58: 674–679; and R. L. Cowie, D. P. Conley, M. F. Underwood, and P. G. Reader. 2008. A randomised controlled trial of the Buteyko technique as an adjunct to conventional management of asthma. *Respiratory Medicine* 102: 726–732.

24. Jane E. Brody. 2009. A Breathing technique offers help for people with asthma. *New York Times*. November 2. Retrieved on October 28, 2017, from http://nyti.ms/28Ns7iV.

25. For details and the famous "Bohr effect" see F. B. Jensen. 2004. Red blood cell pH, the Bohr effect, and other oxygenation-linked phenomena in blood O2 and CO2 transport. *Acta physiologica Scandinavica* 182: 215–227.

26. D. J. Abbott, F. M. Baroody, E. Naureckas, and R. M. Naclerio. 2001. Elevation of nasal mucosal temperature increases the ability of the nose to warm and humidify air. *American Journal of Rhinology* 15: 41–45.

27. D. E. Lieberman. 2011. The evolution of the human head. Harvard University Press.

28. P. McKeown. 2010. *Buteyko meets Dr. Mew*. ButeykoClinic.com.

29. GOPex, a type of oral-facial therapy focused on posture (technically "myopostural"). It is, however, easily mistaken for a well-established type of physiotherapy (PT) similar to speech therapy (technically "myofunctional therapy.") The latter helps to retrain oral muscle memory. It is similar to the rehabilitation needed when you have had an accident and lost the use of a limb. You need then to relearn the muscle memory that you have lost for that limb.

Oral-facial myo*functional* therapy trains children who don't use their face and mouth muscles well, to regain control through physiotherapy, or mouth-swallow tongue therapy. These exercises target function—movement—and can be very effective for the purposes for which they are designed. But they play a minor role in guiding us in our oral-facial growth and development.

30. S. Kahn and S. Wong. 2016. *GOPex: Good oral posture exercises*. Self-published.

31. F. B. Jensen. 2004. Red blood cell pH, the Bohr effect, and other oxygenation-linked phenomena in blood O2 and CO2 transport. *Acta physiologica Scandinavica* 182: 215–227.

32. See, for example, T. R. Belfor. 2014. Airway development through dental appliance therapy. *Journal of Sleep Disorders & Therapy* 3 (178) 2167-0277; D. Mahony and T. Belfor. Anti-Ageing Medicine and Orthodontic Appliance Therapy Treatment: An Interdisciplinary Approach, http://asnanportal.com/images/Orthodontics/ANTI-AGING MEDICINE ORTHODONTIC APPLIANCE.pdf; and G. Singh, J. Diaz, C. Busquets-Vaello, and T. Belfor. 2003. Facial changes following treatment with a removable orthodontic appliance in adults. *The Functional Orthodontist* 21: 18–20, 22–13.

33. Kevin Boyd, personal communication, March 12, 2016.

34. B. Melsen, L. Attina, M. Santuari, and A. Attina. 1987. Relationships between swallowing pattern, mode of respiration, and development of malocclusion. *The Angle Orthodontist* 57: 113–120.

CHAPTER 8

1. Remember also that these statistics are just rough estimates.

2. E. Tausche, O. Luck, and W. Harzer. 2004. Prevalence of malocclusions in the early mixed dentition and orthodontic treatment need. *The European Journal of Orthodontics* 26: 237–244.

3. L. E. J. Johnston. 1999. Growing jaws for fun and profit: A modest proposal. In

Growth modification: What works, what doesn't, and why, J. McNamara Jr., ed.: 63–86. *Twenty-Fifth Annual Moyers Symposium*, vol. 35. Ann Arbor: University of Michigan.

4. Kevin Boyd. 2016. Pre- and post-natal retrognathia in *Homo sapiens*: An evolutionary perspective on a modern, and serious, pediatric health problem, Retrieved on October 28, 2017, from http://bit.ly/2bM4qpA.

5. P. R. Ehrlich. 2000. Human natures: Genes, cultures, and the human prospect. Island Press.

6. R. M. Little. 1999. Stability and relapse of mandibular anterior alignment: University of Washington studies. In *Seminars in orthodontics*: 191–204. Elsevier; R. M. Little, R. A. Riedel, and J. Artun J. 1988. An evaluation of changes in mandibular anterior alignment from 10 to 20 years postretention. *American Journal of Orthodontics and Dentofacial Orthopedics* 93: 423–428; and R. M. Little, T. R. Wallen, and R. A. Riedel. 1981b. Stability and relapse of mandibular anterior alignment: First premolar extraction cases treated by traditional edgewise orthodontics. *American Journal of Orthodontics* 80: 349–365.

7. R. M. Little. 1999. Stability and relapse of mandibular anterior alignment: University of Washington studies. *Seminars in orthodontics*: 191–204. Elsevier.

8. M. Mew. 2009. A black swan? *British Dental Journal* 206: 393–393.

9. J. Alió-Sanz, C. Iglesias-Conde, J. Lorenzo-Pernía, A. Iglesias-Linares, A. Mendoza-Mendoza, and E. Solano-Reina. 2012. Effects on the maxilla and cranial base caused by cervical headgear: A longitudinal study. *Med Oral Patol Oral Cir Bucal* 17: e845–e851.

10. Kirsi Pirilä-Parkkinen, Pertti Pirttiniemi, Peter Nieminen, Heikki Löppönen, Uolevi Tolonen, Ritva Uotila, and Jan Huggare. 1999. Cervical headgear therapy as a factor in obstructive sleep apnea syndrome. *Pediatric Dentistry* 21: 39–45.

11. L. E. J. Johnston. 1999. Growing jaws for fun and profit: A modest proposal. In *Growth modification: What works, what doesn't, and why,* J. McNamara Jr., ed.: 63–86. *Twenty-Fifth Annual Moyers Symposium*, vol. 35. Ann Arbor: University of Michigan.

12. J. R. C. Mew. 2007. Facial changes in identical twins treated by different orthodontic techniques. *World Journal of Orthodontics* 8: 174–187.

13. J. R. C. Mew. 1981. The aetiology of malocclusion: Can the tropic premise assist our understanding? *British Dental Journal* 151: 296–301; and J. R. C. Mew. 1993. Forecasting and monitoring facial growth. *American Journal of Orthodontics and Dentofacial Orthopedics* 104: 105–120.

14. J. McNamara. 1981a. Influence of respiratory pattern on craniofacial growth. *Angle Orthodont.* 51: 269–300,; and J. A. J. McNamara. 1981b. Components of class II malocclusion in children 8–10 years of age. *Angle Orthodont.* 51: 177–202.

15. W. Proffit. 1978. Equilibrium theory revisited: factors influencing position of the teeth. *Angle Orthodont.* 48: 175–185.

16. D. E. Lieberman. 2011. *The evolution of the human head.* Cambridge, MA: Harvard University Press.

17. I. Bondemark, A.-K. Holm , K. Hansen, S. Axelsson, B. Mohlin, V. Brattstrom, G. Paulin, and T. Pietila. 2007. Long-term stability of orthodontic treatment and patient satisfaction: A systematic review. *Angle Orth.* 77: 181–191.

18. R. Little, T. Wallen, and R. Riedel. 1981. Stability and relapse ot mandibular anterior alignment: First premotar extraction cases treated by traditional edgewise orthodontics. *American Journal of Orthodontics* 80: 349–365.

19. J. R. C. Mew. 2007. Facial changes in identical twins treated by different orthodontic techniques. *World Journal of Orthodontics* 8: 174–187.

20. J. R. C. Mew, personal communication.

21. Simon Wong, personal communication,

22. J. R. C. Mew. 2004b. The postural basis of malocclusion: A philosophical overview. *The American Journal of Orthodontics and Dentofacial Orthopedics* 126: 729–738.

23. J. R. C. Mew. 2007. Facial changes in identical twins treated by different orthodontic techniques. *World Journal of Orthodontics* 8: 174–187.

24. Simon Wong, personal communication.

25. Personal communication from Dr. Michael Mew and one of his patients to Sandra.

26. DNA appliance system journal articles. Retrieved on October 28, 2017, from http://bit.ly/1p2mC3g; G. Singh, T. Griffin, and R. Chandrashekhar. 2014. Biomimetic oral appliance therapy in adults with mild to moderate obstructive sleep apnea. *Austin J Sleep Disord* 1: 5; W. Harris and G. Singh. 2013. Resolution of "gummy smile" and anterior open bite using the DNA appliance: Case Report. *J Amer Orthod Soc*: 30–34.

27. L. E. J. Johnston. 1999. Growing jaws for fun and profit: A modest proposal. In *Growth modification: What works, what doesn't, and why*, J. McNamara Jr., ed.: 63–86. *Twenty-Fifth Annual Moyers Symposium*, vol. 35. Ann Arbor: University of Michigan.

28. C. F. Aelbers and L. Dermaut. 1996. Orthopedics in orthodontics: Part I, Fiction or reality: A review of the literature. *American Journal of Orthodontics and Dentofacial Orthopedics* 110: 513–519; and L. Dermaut and C. Aelbers C. 1996. Orthopedics in orthodontics: Fiction or reality. A review of the literature, part II. *American Journal of Orthodontics and Dentofacial Orthopedics* 110: 667–671.

29. P. Agostino, A. Ugolini, A. Signori, A. Silvestrini-Biavati, J. E. Harrison, and P. Riley. 2014. Orthodontic treatment for posterior crossbites. *Cochrane Database Syst Rev* 8; F. R. Borrie, D. R. Bearn, N. Innes, and Z. Iheozor-Ejiofor. 2015. Interventions for the cessation of non-nutritive sucking habits in children. *The Cochrane Database of Systematic Reviews* 3; F. R. Carvalho, D. Lentini-Oliveira, M. Machado, G. Prado, L. Prado, and H. Saconato. 2007. Oral appliances and functional orthopaedic appliances for obstructive sleep apnoea in children. *Cochrane Database Syst Rev* 2; H. Minami-Sugaya, D, A, Lentini-Oliveira, F, R, Carvalho, M. A. C. Machado, C. Marzola, H. Saconato, and G. F. Prado. 2012. Treatments for adults with prominent lower front teeth. The Cochrane Library; N. Parkin, S. Furness, A. Shah, B. Thind, Z. Marshman, G. Glenroy, F. Dyer, and P. E. Benson. 2012. Extraction of primary (baby) teeth for unerupted palatally displaced permanent canine teeth in children. *Cochrane Database Syst Rev* 12; A. A. Shah. 2003. Postretention changes in mandibular crowding: A review of the literature. *American Journal of Orthodontics and Dentofacial Orthopedics* 124: 298–308; B. Thiruvenkatachari, J. E. Harrison, H. V. Worthington, and K. D. O'Brien. 2013. Orthodontic treatment for prominent upper front teeth (Class II malocclusion) in children. *Cochrane Database Syst Rev* 11; S. Watkinson, J. E. Harrison, S. Furness, and H. V. Worthington. 2013. Orthodontic treatment for prominent lower front teeth (Class III malocclusion) in children. The Cochrane Library; and Y. Yu, J. Sun, W. Lai, T. Wu, S. Koshy, and Z. Shi. 2013. Interventions for managing relapse of the lower front teeth after orthodontic treatment. The Cochrane Library.

30. K. B. Kim. 2015. How has our interest in the airway changed over 100 years? *American Journal of Orthodontics and Dentofacial Orthopedics* 148: 740–747.

31. D. M'Kenzie D. 1915. Some points of common interest to the rhinologist and the orthodontist. *International Journal of Orthodontia* 1: 9–17.

32. P. 7.

33. http://aapmd.org/

34. D. H. Enlow and M. G. Hans. 1996. *Essentials of facial growth.* Saunders.

35. N. Stefanovic, H. El, D. L. Chenin, B. Glisic, and J. M. Palomo. 2013. Three-dimensional pharyngeal airway changes in orthodontic patients treated with and without extractions. *Orthod. Craniofac. Res.* 16: 87–96.

36. L. Schropp, A. Wenzel, L. Kostopoulos, and T. Karring. 2003. Bone healing and soft tissue contour changes following single-tooth extraction: a clinical and radiographic 12–month prospective study. *International Journal of Periodontics and Restorative Dentistry* 23: 313–324; F. Van der Weijden, F. Dell'Acqua, and D. E. Slot. 2009. Alveolar bone dimensional changes of post-extraction sockets in humans: A systematic review. *Journal of Clinical Periodontology* 36: 1048–1058.

37. A. J. Larsen, D. B. Rindal, J. P. Hatch, S. Kane, S. E. Asche, C. Carvalho, and J. Rugh. 2015. Evidence supports no relationship between obstructive sleep apnea and premolar extraction: An electronic health records review. Journal of Clinical Sleep Medicine 11: 1443.

38. C. Guilleminault, V. C. Abad, H.-Y. Chiu, B. Peters, and S. Quo. 2016. Missing teeth and pediatric obstructive sleep apnea. *Sleep and Breathing* 20: 561–568; and B. H. Seto, H. Gotsopoulos, M. R. Sims, and P. A. Cistulli PA. 2001. Maxillary morphology in obstructive sleep apnoea syndrome. *The European Journal of Orthodontics* 23: 703–714.

39. L. E. J. Johnston. 1999. Growing jaws for fun and profit: A modest proposal. In *Growth modification: What works, what doesn't, and why,* J. McNamara Jr., ed.: 63–86. *Twenty-fifth Annual Moyers Symposium,* vol. 35. Ann Arbor: University of Michigan.

40. J. W. Friedman. 2007. The prophylactic extraction of third molars: A public health hazard. *Am J Public Health* 97: 1554–1559; J. W. Friedman. 2008. Friedman responds. *American Journal of Public Health* 98: 582.

CHAPTER 9

1. L. G. Abreu, S. M. Paiva, I. A. Pordeus, and C. C. Martins. 2016. Breastfeeding, bottle feeding and risk of malocclusion in mixed and permanent dentitions: A systematic review. *Brazilian Oral Research* 30; J.-L. Raymond. 2000. A functional approach to the relationship between nursing and malocclusion. *Revue D'Orthopedie Dentofaciale* 34: 379–404; and J. Raymond and W. Bacon. 2006. Influence of feeding method on maxillofacial development. *L'Orthodontie française* 77: 101–103.

2. S. A. S. Moimaz, A. J. Í. Garbin, A. M. C. Lima, L. F. Lolli, O. Saliba, and C. A. S. Garbin. 2014. Longitudinal study of habits leading to malocclusion development in childhood. *BMC Oral Health* 14: 96.

3. U. Deb and S. N. Bandyopadhyay. 2007. Care of nasal airway to prevent orthodontic problems in children. *J. Indian Med Assoc.* 105: 640, 642; D. Harari, M. Redlich, S. Miri, T. Hamud, and M. Gross 2010. The effect of mouth breathing versus nasal breathing on dentofacial and craniofacial development in orthodontic patients. *Laryngoscope* 120: 2089–2093; and S. E. Mattar, W. T. Anselmo-Lima, F. C. Valera, and M. A. Matsumoto. 2004b. Skeletal and occlusal characteristics in mouth-breathing pre-school children. *J Clin Pediatr Dent.* 28: 315–318.

4. F. W. Booth, S. E. Gordon, C. J. Carlson, and M. T. Hamilton. 2000. Waging war on modern chronic diseases: Primary prevention through exercise biology. *Journal of Applied Physiology* 88: 774–787.

5. T. Bodenheimer, E. Chen, and H. D. Bennett. 2009. Confronting the growing burden of chronic disease: Can the US health care workforce do the job? *Health Affairs* 28: 64–74.

6. Ad.-G. Aikins, N. Unwin, C. Agyemang, P. Allotey, C. Campbell, and D. Arhinful. 2010. Tackling Africa's chronic disease burden: From the local to the global. *Globalization and Health* 6: 1.

7. R. Sapolsky. 1997. *The trouble with testosterone and other essays on the biology of the human predicament.* Scribner.

8. P. Dasgupta and P. R. Ehrlich. 2013. Pervasive externalities at the population, consumption, and environment nexus. *Science* 340: 324–328; P. R. Ehrlich and A. H. Ehrlich. 2013. Can a collapse of civilization be avoided? *Proceeding of the Royal Society B. Available at* http://rspb.royalsocietypublishing.org/content/280/1754/20122845; P. R. Ehrlich and J. Harte. 2015a. Food security requires a new revolution. *International Journal of Environmental Studies. Available at* http://dx.doi.org/10.1080/00207233.2015.1067468:1–13; P. R. Ehrlich, P. M. Kareiva, and G. C. Daily. 2012. Securing natural capital and expanding equity to rescale civilization. *Nature*, 486: 68–73; and J. Harte. 2007. Human population as a dynamic factor in environmental degradation. *Population and Environment* 28: 223–236.

9. P. R. Ehrlich and M. W. Feldman. 2003. Genes and cultures: What creates our behavioral phenome? *Current Anthropology* 44: 87–107.

10. D. Perlmutter. 2013. Grain brain: The surprising truth about wheat, carbs, and sugar—your brain's silent killers. Little, Brown, and Company.

11. M. Klatsky and R. L. Fisher. 1953. *The human masticatory apparatus: An introduction to dental anthropology.* Dental Items of Interest Pub. Co.

12. P, J, Brekhus. 1941. *Your teeth, their past, present, and probable future.* University of Minnesota Press; and R. S. Corruccini. 1999. How anthropology informs the orthodontic diagnosis of malocclusion's causes. Edwin Mellen Press.

13. E. Touchette. 2011. Factors associated with sleep problems in early childhood. *Encyclopedia on Early Childhood Development* March: 1–8.

14. P. Defabjanis. 2004. Impact of nasal airway obstruction on dentofacial development and sleep disturbances in children: Preliminary notes. *Journal of Clinical Pediatric Dentistry* 27: 95–100.

15. A. Sheiham and R. G. Watt. 2000. The common risk factor approach: A rational basis for promoting oral health. *Community Dentistry and Oral Epidemiology* 28: 399–406.

16. National Conference of State Legislatures, 2017. Breastfeeding State Laws and Federal Health Reform and Nursing Mothers. June 5. Retrieved on December 4, 2017, from http://bit.ly/1lHJI8E.

17. S. S. Hawkins, A. D. Stern, and M. W. Gillman . 2012. Do state breastfeeding laws in the USA promote breast feeding? *Journal of Epidemiology and Community Health*: jech-2012-201619.

18. C. Parcells, M. Stommel, and R, P, Hubbard. 1999. Mismatch of classroom furniture and student body dimensions: Empirical findings and health implications. *Journal of Adolescent Health* 24: 265–273.

19. Ibid.

20. J. Cawley. 2010. The economics of childhood obesity. *Health Affairs* 29: 364–371; and A. Freeman. 2007. Fast food: Oppression through poor nutrition. *California Law Review* 95.

21. K. D. Brownell, R. Kersh, D. S. Ludwig, R. C. Post, R. M. Puhl, M. B. Schwartz, and W. C. Willett. 2010. Personal responsibility and obesity: A constructive approach to a controversial issue. *Health Affairs* 29: 379–387.

22. N. A. Christakis and J. H. Fowler. 2007. The spread of obesity in a large social network over 32 years. *New England Journal of Medicine* 357: 370–379.

23. H. Bruckner and P. Bearman. 2005. After the promise: The STD consequences of adolescent virginity pledges. *J Adolesc Health* 36: 271–278.

24. G. H. Montgomery, J. Erblich, T. DiLorenzo, and J. H. Bovbjerg. 2003. Family and friends with disease: Their impact on perceived risk. *Preventive Medicine* 37: 242–249.

25. J. A. Bernstein, N. Alexis, H. Bacchus, I. L. Bernstein, P. Fritz, E. Horner, N. Li, S. Mason, A. Nel, and J. Oullette. 2008. The health effects of nonindustrial indoor air pollution. *Journal of Allergy and Clinical Immunology* 121: 585–591.

26. P. R. Ehrlich and J. Harte. 2015a. Food security requires a new revolution. *International Journal of Environmental Studies:* 1–13. Available at http://dx.doi.org/10.1080/0020 7233.2015.1067468; and P. R. Ehrlich and J. Harte. 2015b. Opinion: To feed the world in 2050 will require a global revolution. *Proc Natl Acad Sci USA* 112:14743–14744.

27. D. Tilman, C. Balzer, J. Hill, and B. L. Befort. 2011. Global food demand and the sustainable intensification of agriculture. *Proc Natl Acad Sci USA* 108:20260–20264; and D. Tilman and M. Clark. 2014. Global diets link environmental sustainability and human health. *Nature* 515: 518–522.

ABOUT THE AUTHORS

Sandra Kahn, DDS, MSD, a Diplomate of the American Board of Orthodontics, has 22 years of clinical experience treating thousands of patients, almost all children. She has served on craniofacial anomalies teams, both at Stanford University and University of California in San Francisco, teaches, and lectures internationally. Over years of practice using standard orthodontic techniques, she grew increasingly frustrated. Although the "results" were satisfactory by the standards of the profession, they were temporary unless a retainer held the teeth in their new positions. Like other orthodontists, she was seeing remissions, not cures. When her son Ilan began to show signs of problems with his airway—snoring and mouth-open posture—she began to explore an alternative orthodontic method known as orthotropics (meaning "straight growth"). Orthotropics enabled the development of teeth that are not crowded while encouraging facial development in a way that increases the size of the airway and in turn prevents snoring and sleep apnea. Today she focuses her practice on forwardontics, with a concentration on pediatric prevention of long-term obstructive sleep apnea.

Paul R. Ehrlich is Bing Professor of Population Studies Emeritus and President of the Center for Conservation Biology at Stanford University. His research on population genetics, genetic and cultural evolution, population, and environment has resulted in some 50 books and more than 1,000 articles. He has pursued field experiments and programs of observation of the behavior and ecology on animals as different as butterflies, birds, mammals, reef fishes, mites, and human beings. He is a cofounder of the discipline of coevolution and is probably best known for his many analyses of the human predicament and, especially, the roles overpopulation, overconsumption, and inequity play in it. In his spare time he was a correspondent for NBC News and, later, among his thousands of media appearances he performed some 20 times on the *Tonight Show* with Johnny Carson. His honors include fellowship in the United States National Academy of Sciences and the American Academy of Arts and Sciences; membership in the American Philosophical Society; foreign member of the Royal Society of London; Honorary Fellowship, Royal Entomological Society of London; the Eminent Ecologist Award, Ecological Society of America; California Academy of Sciences, Fellows' Medal; Fellow, Beijer Institute, Royal Swedish Academy of Sciences; Margalef Prize in Ecology and Environmental Sciences; Tyler Prize for Environmental Achievement; Dr. A. H. Heineken Prize for Environmental Sciences; National

Audubon Society, One Hundred Champions of Conservation; Volvo Environment Prize; International Center for Tropical Ecology, World Ecology Medal; United Nations Environment Programme Sasakawa Environment Prize; Heinz Award for the Environment; Nuclear Age Peace Foundation, Distinguished Peace Leader; Blue Planet Prize of the Asahi Glass Foundation, Japan; AAAS/Scientific American Award for Service to Science in the Cause of Humankind; United Nations Global 500 Roll of Honour; Honorary Life Member, American Humanist Association; Distinguished Scientist Award, American Institute of Biological Sciences; and the Crafoord Prize in Population Biology and Conservation of Biological Diversity, The Royal Swedish Academy of Sciences (explicit substitute for the Nobel Prize in fields in which the Nobel is not given).

Paul and Sandra's connections go beyond the pleasures of dining with their partners. Sandra Kahn and David Leventhal got to know Paul and Anne Ehrlich because of their mutual interest in saving the animals, plants, and microbes with which we all share Earth. Sandra and David had founded a conservation organization, RAINFOREST 2 REEF; Paul and Anne, as ecologists/evolutionists and conservation biologists were introduced to Sandra and David by a colleague. Mutual interests soon metamorphosed into a family friendship, and the dining program already mentioned. This volume is thus the result of a friendship, convergent interests, and recognition by four friends that there is a great and generally unrecognized need to spare multitudes of children from having less-than-satisfactory futures.

INDEX

Note: Page numbers in italic type indicate illustrations.

CPSIA information can be obtained
at www.ICGtesting.com
Printed in the USA
JSHW050152101121
20329JS00009B/183